CLEP

College Level Examination Program

Biology

Jeffery Sack, Ed.D

XAMonline

Copyright © 2016

All rights reserved. No part of the material protected by this copyright notice may be reproduced or utilized in any form or by any means, electronic or mechanical, including photocopying or recording or by any information storage and retrievable system, without written permission from the copyright holder.

To obtain permission(s) to use the material from this work for any purpose including workshops or seminars, please submit a written request to:

XAMonline, Inc.
21 Orient Avenue
Melrose, MA 02176
Toll Free: 1-800-301-4647
Email: info@xamonline.com
Web: www.xamonline.com
Fax: 1-617-583-5552

Library of Congress Cataloging-in-Publication Data
Sack, Jeffrey

CLEP Biology/ Jeffrey Sack
 ISBN: 978-1-60787-531-4

1. CLEP 2. Study Guides 3. Biology

Disclaimer:

The opinions expressed in this publication are the sole works of XAMonline and were created independently from The College Board, or other testing affiliates. Between the time of publication and printing, specific test standards as well as testing formats and website information may change that are not included in part or in whole within this product. XAMonline develops sample test questions, and they reflect similar content as on real tests; however, they are not former tests. XAMonline assembles content that aligns with test standards but makes no claims nor guarantees candidates a passing score.

© Can Stock Photo Inc./kasto/14163272

Printed in the United States of America
CLEP Biology
ISBN: 978-1-60787-531-4

Table of Contents

Chapter 1: All of the Details . 1

 The College-Level Examination Program
 How the Program Works . 1
 The CLEP Examinations . 2
 What the Examinations Are Like . 2
 Where to Take the Examinations and How to Register 3
 ACE's College Credit Recommendation Service 3
 How Your Score Is Reported . 4
 Approaching a College about CLEP . 4
 How to Apply for College Credit . 5
 Questions to Ask about a College's CLEP Policy 8
 Preparing to Take CLEP Examinations 10
 Test Preparation Tips . 10
 Accommodations for Students with Disabilities 12
 Interpreting Your Scores . 13
 How CLEP Scores Are Computed . 13
 How Essays Are Scored . 14
 The CLEP Biology Exam . 14
 Knowledge and Skills Required . 15

Chapter 2: How Science is Done . 19

 The Scientific Method . 19
 State the Problem . 20
 Collect Background Information . 20
 Establish a Hypothesis . 21
 Perform the Experiment . 21
 Analyze the Data . 23
 Repeat the Experiment . 24
 Draw Conclusions . 24
 Report the Results . 25
 The Scientific Process . 25
 Scientific Facts . 26
 Scientific Concepts . 26
 Scientific Models . 26
 Biology Does Math Too! . 27
 Measurement and Notation Systems . 27
 How to Manipulate Your Data . 28

Commonly Shared Scientific Ideals 29
How the Scientific Method Works Outside of Science 30

Chapter 3: Cellular Chemistry 33

The Chemistry of Biology 33
Atomic Structure..................................... 33
Compounds, Formulas, and Bonding – Oh My! 35
 pH and Buffers................................... 37
 Water. It's What Makes Life Possible. 38
 The Molecules of Life. 39
 Functional Groups 40
 Carbohydrates 41
 Lipids ... 42
 Proteins ... 43
 Enzymes .. 45
 Factors Affecting Enzymes 45
 Cofactors and Coenzymes 46
 Nucleic Acids 46
DNA Replication..................................... 47
 Did I Make a Mistake?............................. 49

Chapter 4: Energy of the Cell 51

Energy and Thermodynamics 51
Cells and Energy: The Two Big Processes 52
 Redox Reactions 52
Photosynthesis...................................... 53
 Photosynthetic Pigments 54
 How It Happens 55
 Variations on a Theme 57
Cellular Respiration 58
 Glycolysis.. 59
 Kreb's Cycle..................................... 59
 Electron Transport Chain 61
Aerobic versus Anaerobic Respiration 62

Chapter 5: Cell Structure and Function............... 65

Cell Diversity....................................... 65
 Prokaryotic Cells 66
 Eukaryotic Cells.................................. 67
Organelles.. 68

Getting Into and Out of the Cell. 72
 Cell Membrane. 72
Put Your Left Foot In. Take Your Left Foot Out 73
Cell Transport. 74
 Osmotic Potential. 74
Active Transport. 75
How Cells Divide: Mitosis and Meiosis. 76
 Mitosis . 76
 Meiosis. 79

Chapter 6: Genetics. 81

Mendel's Laws . 81
Exceptions to the Rule. 86
Family History. 87
More Exceptions to the Rule . 87
 Multiple Genes. 88
 Organelle Genetics . 88
 Transmission Bias . 88
Genes Can Be Linked. 88
Human Genetic Disorders . 89
 Sex influenced traits. 90
 Chromosome Theory. 90
 Screening for Genetic Disorders . 90
Impact of Mutations . 92
Genes and the Environment . 92

Chapter 7: Evolution . 93

DNA and Molecular Similarities. 93
Body Structures . 94
Similarities in Embryological Development. 95
Plate Tectonics and Biogeography. 96
Fossil Record . 96
The Ideas of Evolution . 96
 Lamarck and "Use and Disuse" . 97
 Darwin and Natural Selection . 97
 Types of Natural Selection. 99
 Changes in Allele Frequency. 99
 How Do New Species Form? . 100
Patterns of Evolution . 100
Different Ideas about Evolution. 101
From Life to Death . 102

Chapter 8: Diversity of Life 105

The Science of Classifying Living Things 105
- Binomial Nomenclature 106
- Does King Phillip Come Over For Good Science? 106

The Diversity of Life – An Overview 109
- Viruses .. 109
- Domain Bacteria 110
- Domain Archaea 111
- Domain Eukarya 111
- Kingdom Protista 112
- Kingdom Fungi 113
- Kingdom Plantae 114
 - *Plant evolution and diversity* 115
- Kingdom Animalia 118
 - *Animal evolution and diversity* 118
 - *Survey of the Animal Kingdom* 119

Evolutionary History 122

Chapter 9: Organismal Biology 123

Plant structure and function 123
- Roots ... 123
- Stems ... 124
- Leaves .. 125
- Hormones Drive Plant Behavior 126
 - *Tropisms* 127
 - *Photoperiods* 127

Animal Structure and Function 127
- Digestion 128
 - *Complex Animal Digestion* 128
- Circulation 129
 - *Complex Animal Circulation* 130
- Respiration 131
 - *Complex Animal Respiration* 132
- Excretion 132
 - *Complex Animal Excretion* 133
- Structure, Support, and Protection 134
- Let's Get Moving 134
- Information Transmission 136
 - *Complex Organism Signal Transfer* 136
 - *Divisions of the Nervous System* 137

 Endocrine System 137
 Hormones in Complex Organisms 138
 The Immune System 139
 Defense in Complex Organisms 140
 Reproduction and Development **142**
 Reproduction and Development in Non-vascular Plants 142
 Reproduction and Development in Vascular Plants 142
 Alternation of generations 142
 Reproduction and Development in Animals 145
 Asexual Reproduction Strategies 145
 Sexual Reproduction Strategies 146
 Development 146
 Reproduction and Development in Complex Organisms . 147
 Animal Behavior **148**
 Animal Social Behavior 149

Chapter 10: Ecology 151

 Energy Gets Its Groove On **151**
 Biological Magnification **153**
 Round and Round They Go – Nutrient Cycles **153**
 Populations **154**
 Communities **157**
 Niches ... 157
 Interspecific relationships **158**
 Ecosystems and biomes **159**
 Desert ... 159
 Grassland .. 159
 Tundra .. 160
 Forests .. 160
 Boreal Forest (Taiga) 160
 Tropical Rainforest 161
 Temperate Forest 161
 Aquatic Ecosystems 161
 Ponds and Lakes 162
 Rivers and Streams 162
 Wetlands 162
 Oceans .. 162
 Coral Reefs 163
 Estuaries 163
 Succession **163**

People are Doing Bad Things: Human Impact on the Planet 164
 You Don't Belong Here – Invasive Species 164
 Resources and their Effects. 165

CLEP Biology Sample Exam 1 . 167

CLEP Biology Test Sample Exam 1: Answer Key and Explanations. 206

CLEP Biology Sample Exam 2 . 265

CLEP Biology Test Sample Exam 2: Answer Key and Explanations. 310

About the Author

Jeffrey Sack

Jeffrey Sack, Ed.D. is a biologist, educator, and writer who has taught all aspects of high school biology in both public and private schools. His scientific interests include marine ecology and bird behavior, and his educational interests include the relationship between teacher scientific content knowledge and pedagogy and how students learn science. Jeff lives in Connecticut, U.S.A. and serves on committees for the National Association of Biology Teachers and the National Marine Educators Association.

Chapter 1: All of the Details

The College-Level Examination Program

CLEP exams are administered at over 1,800 institutions nationwide, and 2,900 colleges and universities award college credit to those who perform well on them. This rigorous program allows many self-directed students of a wide range of ages and backgrounds to demonstrate their mastery of introductory college-level material and pursue greater academic success. Students can earn credit for what they already know by getting qualifying scores on any of the 33 examinations.

How the Program Works

The CLEP exams cover material that is taught in introductory-level courses at many colleges and universities. Faculty at individual colleges review the exams to ensure that they cover the important material currently taught in their courses.

Although CLEP is sponsored by the College Board, only colleges may grant credit toward a degree. To learn about a particular college's CLEP policy, contact the college directly. When you take a CLEP exam, you can request that a copy of your score report be sent to the college you are attending or planning to attend. After evaluating your score, the college will decide whether or not to award you credit for a certain course or courses, or to exempt you from them.

If the college decides to give you credit, it will record the number of credits on your permanent record, thereby indicating that you have completed work equivalent to a course in that subject. If the college decides to grant exemption without giving you credit for a course, you will be permitted to omit a course that would normally be required of you and to take a course of your choice instead.

The CLEP program has a long-standing policy that an exam may not be taken within the specified wait period. This waiting period provides you with an opportunity to spend additional time preparing for the exam or the option of taking a classroom course. If you violate the CLEP retest policy, the administration will be considered invalid, the score canceled, and any test fees will be forfeited. If you are a military service member, please note that DANTES will not fund retesting on a previously funded CLEP exam. However, you may personally fund a retest after the specified wait period.

The CLEP Examinations

CLEP exams cover material directly related to specific undergraduate courses taught during a student's first two years in college. The courses may be offered for three, four, six or eight semester hours in general areas such as mathematics, history, social sciences, English composition, natural sciences and humanities. Institutions will either grant credit for a specific course based on a satisfactory score on the related exam, or in the general area in which a satisfactory is earned. The credit is equal to the credit awarded to students who successfully complete the courses. See the Table of Contents for a complete list of all exam titles.

What the Examinations Are Like

CLEP exams are administered on computer and are approximately 90 minutes long, with the exception of College Composition, which is approximately 120 minutes long. Most questions are multiple-choice; other types of questions require you to fill in a numeric answer, to shade areas of an object, or to put items in the correct order. Questions using these kinds of skills are called zone, shade, grid, scale, fraction, numeric entry, histogram and order match questions.

CLEP College Composition includes a mandatory essay section, responses to which must be typed into the computer.

Some of the examinations have optional essays. You should check with the individual college or university where you are sending your score to see whether an optional essay is required for those exams. These essays are administered on paper and are scored by faculty at the institution that receives your score.

Where to Take the Examinations and How to Register

CLEP exams are administered throughout the year at over 1,800 test centers in the United States and select international sites. Once you have decided to take a CLEP examination, you can log into My Account at https://clepportal.collegeboard.org/myaccount to create and manage your own personal accounts, pay for CLEP exams and purchase study materials. You can self-register at any time by completing the online registration form.

Through My Account you can also access a list of institutions that administer CLEP and locate a test center in your area. **After paying for your exam through My Account, you must still contact the test center to schedule your CLEP exam.**

If you are unable to locate a test center near you, call 800-257-9558 for more information.

ACE's College Credit Recommendation Service

The College Credit Recommendation Service (CREDIT) of the American Council on Education (ACE) enables you to put all of your educational achievements on a secure and universally accepted ACE transcript. All of your ACE-evaluated courses and examinations, including CLEP, appear in an easy-to-read format that includes ACE credit recommendations, descriptions and suggested transfer areas. The service is perfect for candidates who have acquired college credit at multiple ACE-evaluated organizations or credit-by-examination programs. You may have your transcript released at any time to the college of your choice. There is a one-time setup fee of $40 (includes the cost of your first transcript) and a fee of $15 for each transcript requested after release of the first. ACE has an additional transcript service for organizations offering continuing education units.

The College Credit Recommendation Service is offered through ACE's Center for Lifelong Learning. For more than 50 years, ACE has been at the forefront of the evaluation of education and training attained outside the classroom. For more information about ACE CREDIT, contact:

ACE CREDIT
One Dupont Circle, NW
Suite 250
Washington, DC 20036

ACE's Call Center is open Monday to Friday, 8:45 a.m. to 4:45 p.m., and can be reached at 866-205-6267 or CREDIT@ace.nche.edu. Staff are able to assist you with courses and certifications that carry ACE recommendations for both civilian organizations and training obtained through the military.

If you are already registered for an ACE transcript, you can access your records and order transcripts using the ACE Online Transcript System: https://www.acenet.edu/transcripts/.

ACE's Center for Lifelong Learning can be found on the Internet at: http://www.acenet.edu/higher-education.

How Your Score Is Reported

You have the option of seeing your CLEP score immediately after you complete the exam, except in the case of College Composition, for which scores are available four to six weeks after the exam date. Once you choose to see your score, it will be sent automatically to the institution you have designated as a score recipient; it cannot be canceled. You will receive a candidate copy of your score before you leave the test center. If you have tested at the institution that you have designated as a score recipient, it will have immediate access to your test results.

If you do not want your score reported, you may select that as an option at the end of the examination before the exam is scored. Once you have selected the option to not view your score, the score is canceled.

The score will not be reported to the institution you have designated, and you will not receive a candidate copy of your score report. You will have to wait the specified wait period before you can take the exam again.

CLEP scores are kept on file for 20 years. During this period, for a small fee, you may have your transcript sent to another college or to anyone else you specify. Your score(s) will never be sent to anyone without your approval.

Approaching a College about CLEP

The following sections provide a step-by-step guide to learning about the CLEP policy at a particular college or university. The person or office that can best assist you may have a different title at each institution, but the following guidelines will lead you to information about CLEP at any institution.

Adults and other nontraditional students returning to college often benefit from special assistance when they approach a college. Opportunities for adults to return to formal learning in the classroom are now widespread, and colleges

and universities have worked hard to make this a smooth process for older students. Many colleges have established special offices that are staffed with trained professionals who understand the kinds of problems facing adults returning to college. If you think you might benefit from such assistance, be sure to find out whether these services are available at your college.

How to Apply for College Credit

Step 1. Obtain, or access online, the general information catalog and a copy of the CLEP policy from each college you are considering.

Information about admission and CLEP policies can be obtained on the college's website at clep.collegeboard.org/search/colleges, or by contacting or visiting the admissions office. Ask for a copy of the publication in which the college's complete CLEP policy is explained. Also, get the name and the telephone number of the person to contact in case you have further questions about CLEP.

Step 2. If you have not already been admitted to a college that you are considering, look at its admission requirements for undergraduate students to see whether you qualify.

Whether you're applying for college admission as a high school student, transfer student or as an adult resuming a college career or going to college for the first time, you should be familiar with the requirements for admission at the schools you are considering. If you are a nontraditional student, be sure to check whether the school has separate admissions requirements that might apply to you. Some schools are very selective, while others are "open admission."

It might be helpful for you to contact the admissions office for an interview with a counselor. State why you want the interview and ask what documents you should bring with you or send in advance. (These materials may include a high school transcript, transcript of previous college work or completed application for admission.) Make an extra effort to have all the information requested in time for the interview.

During the interview, relax and be yourself. Be prepared to state honestly why you think you are ready and able to do college work. If you have already taken CLEP exams and scored high enough to earn credit, you have shown that you are able to do college work. Mention this achievement to the admissions counselor because it may increase your chances of being accepted. If you have not taken a CLEP exam, you can still improve your chances of

being accepted by describing how your job training or independent study has helped prepare you for college-level work. Discuss with the counselor what you have learned from your work and personal experiences.

Step 3. Evaluate the college's CLEP policy.

Typically, a college lists all its academic policies, including CLEP policies, in its general catalog or on its website. You will probably find the CLEP policy statement under a heading such as Credit-by-Examination, Advanced Standing, Advanced Placement or External Degree Program. These sections can usually be found in the front of the catalog. You can also check out the institution's CLEP Policy by visiting clep.collegeboard.org/search/colleges.

Many colleges publish their credit-by-examination policies in separate brochures, which are distributed through the campus testing office, counseling center, admissions office or registrar's office. If you find a very general policy statement in the college catalog, seek clarification from one of these offices.

Review the material in the section of this chapter entitled "Questions to Ask about a College's CLEP Policy." Use these guidelines to evaluate the college's CLEP policy. If you have not yet taken a CLEP exam, this evaluation will help you decide which exams to take. Because individual colleges have different CLEP policies, a review of several policies may help you decide which college to attend.

Step 4. If you have not yet applied for admission, do so as early as possible.

Most colleges expect you to apply for admission several months before you enroll, and it is essential that you meet the published application deadlines. It takes time to process your application for admission. If you have yet to take a CLEP exam, you may want to take one or more CLEP exams while you are waiting for your application to be processed. Be sure to check the college's CLEP policy beforehand so that you are taking exams your college will accept for credit. You should also find out from the college when to submit your CLEP score(s).

Complete all forms and include all documents requested with your application(s) for admission.

Normally, an admission decision cannot be reached until all documents have been submitted and evaluated. Unless told to do so, do not send your CLEP score(s) until you have been officially admitted.

Step 5. Arrange to take CLEP exam(s) or to submit your CLEP score(s).

CLEP exams can be taken at any of the 1,800 test centers world-wide. To locate a test center near you. clep.collegeboard.org/search/test-centers.

If you have already taken a CLEP exam, but did not have your score sent to your college, you can have an official transcript sent at any time for a small fee. Fill out the Transcript Request Form included on the same page as your exam score. If you do not have the form, visit clep.collegeboard.org/about/score to download a copy, or call 800-257-9558 to order a transcript using a major credit card. Completed forms should be faxed to 610-628-3726 or sent to the following address, along with a check or money order made payable to CLEP for $20 (this fee is subject to change).

CLEP Transcript Service
P.O. Box 6600
Princeton, NJ 08541-6600

Transcripts will only include CLEP scores for the past 20 years; scores more than 20 years old are not kept on file.

Your CLEP scores will be evaluated, probably by someone in the admissions office, and sent to the registrar's office to be posted on your permanent record once you are enrolled. Procedures vary from college to college, but the process usually begins in the admissions office.

Step 6. Ask to receive a written notice of the credit you receive for your CLEP score(s).

A written notice may save you problems later, when you submit your degree plan or file for graduation. In the event that there is a question about whether or not you earned CLEP credit, you will have an official record of what credit was awarded. You may also need this verification of course credit if you meet with an academic adviser before the credit is posted on your permanent record.

Step 7. Before you register for courses, seek academic advising.

A discussion with your academic adviser can help you to avoid taking unnecessary courses and can tell you specifically what your CLEP credit will mean to you. This step may be accomplished at the time you enroll. Most colleges have orientation sessions for new students prior to each enrollment period. During orientation, students are usually assigned academic advisers who then give them individual help in developing long-range plans and course schedules for the next semester. In conjunction with this counseling, you may be asked to take some additional tests so that you can be placed at the proper course level.

Questions to Ask about a College's CLEP Policy

Before taking CLEP exams for the purpose of earning college credit, try to find the answers to these questions:

1. Which CLEP exams are accepted by the college?

 A college may accept some CLEP exams for credit and not others — possibly not the exams you are considering. For this reason, it is important that you know the specific CLEP exams for which you can receive credit.

2. Does the college require the optional free-response (essay) section for exams in composition and literature as well as the multiple-choice portion of the CLEP exam you are considering? Will you be required to pass a departmental test such as an essay, laboratory or oral exam in addition to the CLEP multiple-choice exam?

 Knowing the answers to these questions ahead of time will permit you to schedule the optional free-response or departmental exam when you register to take your CLEP exam.

3. Is CLEP credit granted for specific courses at the college? If so, which ones?

 You are likely to find that credit is granted for specific courses and that the course titles are designated in the college's CLEP policy. It is not necessary, however, that credit be granted for a specific course for you to benefit from your CLEP credit. For instance, at many liberal arts colleges, all students must take certain types of courses; these courses may be labeled the core curriculum, general education requirements, distribution requirements or liberal arts requirements. The requirements are often expressed in terms of credit hours. For example, all students may be required to take at least six hours of humanities, six hours of English, three hours of mathematics, six hours of natural science and six hours of social science, with no particular courses in these disciplines specified. In these instances, CLEP credit may be given as "6 hrs. English Credit" or "3 hrs. Math Credit" without specifying for which English or mathematics courses credit has been awarded. To avoid possible disappointment, you should know before taking a CLEP exam what type of credit you can receive or whether you will be exempted from a required course but receive no credit.

4. How much credit is granted for each exam you are considering, and does the college place a limit 0n the total amount of CLEP credit you can earn toward your degree?

 Not all colleges that grant CLEP credit award the same amount for individual exams. Furthermore, some colleges place a limit on the total amount of credit you can earn through CLEP or other exams. Other colleges may grant you exemption but no credit toward your degree. Knowing several colleges' policies concerning these issues may help you decide which college to attend. If you think you are capable of passing a number of CLEP exams, you may want to attend a college that will allow you to earn credit for all or most of them. Check out if your institution grants CLEP policy by visiting clep.collegeboard.org/search/colleges.

5. What is the required score for earning CLEP credit for each exam you are considering?

 Most colleges publish the required scores for earning CLEP credit in their general catalogs or in brochures. The required score may vary from exam to exam, so find out the required score for each exam you are considering.

6. What is the college's policy regarding prior course work in the subject in which you are considering taking a CLEP exam?

 Some colleges will not grant credit for a CLEP exam if the candidate has already attempted a college-level course closely aligned with that exam. For example, if you successfully completed English 101 or a comparable course on another campus, you will probably not be permitted to also receive CLEP credit in that subject. Some colleges will not permit you to earn CLEP credit for a course that you failed.

7. Does the college make additional stipulations before credit will be granted?

 It is common practice for colleges to award CLEP credit only to their enrolled students. There are other stipulations, however, that vary from college to college. For example, does the college require you to formally apply for or to accept CLEP credit by completing and signing a form? Or does the college require you to "validate" your

CLEP score by successfully completing a more advanced course in the subject? Getting answers to these and other questions will help to smooth the process of earning college credit through CLEP.

Preparing to Take CLEP Examinations

Test Preparation Tips

1. Familiarize yourself as much as possible with the test and the test situation before the day of the exam. It will be helpful for you to know ahead of time
 a. how much time will be allowed for the test and whether there are timed subsections. (This information is included in the examination guides and in the CLEP Tutorial video.)
 b. what types of questions and directions appear on the exam. (See the examination guides.)
 c. how your test score will be computed.
 d. in which building and room the exam will be administered.
 e. the time of the test administration.
 f. direction, transit and parking information to the test center.
2. Register and pay your exam fee through My Account at https://clepportal.collegeboard.org/myaccount and print your registration ticket. Contact your preferred test center to schedule your appointment to test. Your test center may require an additional administration fee. Check with your test center and confirm the amount required and acceptable method of payment.
3. On the day of the exam, remember to do the following:
 a. Arrive early enough so that you can find a parking place, locate the test center, and get settled comfortably before testing begins.
 b. Bring the following with you:
 o completed registration ticket
 o any registration forms or printouts required by the test center. Make sure you have filled out all necessary paperwork in advance of your testing date.
 o a form of valid and acceptable identification. Acceptable identification must:
 ▪ Be government-issued

- Be an original document — photocopied documents are not acceptable
- Be valid and current — expired documents (bearing expiration dates that have passed) are not acceptable, no matter how recently they may have expired
- Bear the test-taker's full name, in English language characters, exactly as it appears on the
- Registration Ticket, including the order of the names.
- Middle initials are optional and only need to match the first letter of the middle name when present on both the ticket and the identification.
- Bear a recent recognizable photograph that clearly matches the test-taker
- Include the test-taker's signature
- Be in good condition, with clearly legible text and a clearly visible photograph

Refer to the Exam Day Info page on the CLEP website (http://clep.collegeboard.org/exam-day-info) for more details on acceptable and unacceptable forms of identification.

- o military test-takers, bring your Geneva Convention Identification Card. Refer to clep.collegeboard.org/military for additional information on IDs for active duty members, spouses, and civil service civilian employees.
- o two number 2 pencils with good erasers. Mechanical pencils are prohibited in the testing room.

c. Leave all books, papers and notes outside the test center. You will not be permitted to use your own scratch paper; it will be provided by the test center.

d. Do not take a calculator to the exam. If a calculator is required, it will be built into the testing software and available to you on the computer. The CLEP Tutorial video will have a demonstration on how to use online calculators.

e. Do not bring a cell phone or other electronic devices into the testing room.

4. When you enter the test room, be prepared.
 a. You will be assigned to a computer testing station. If you have special needs, be sure to communicate them to the test center administrator before the day you test.
 b. Be relaxed while you are taking the exam. Read directions carefully and listen to all instructions given by the test administrator. If you don't understand the directions, ask for help before the test begins. If you must ask a question that is not related to the exam after testing has begun, raise your hand and a proctor will assist you. The proctor cannot answer questions related to the exam.
 c. Know your rights as a test-taker. You can expect to be given the full working time allowed for taking the exam and a reasonably quiet and comfortable place in which to work. If a poor testing situation is preventing you from doing your best, ask whether the situation can be remedied. If it can't, ask the test administrator to report the problem on a Center Problem Report that will be submitted with your test results. You may also wish to immediately write a letter to CLEP, P.O. Box 6656, Princeton, NJ 08541- 6656. Describe the exact circumstances as completely as you can. Be sure to include the name of the test center, the test date and the name(s) of the exam(s) you took.

Accommodations for Students with Disabilities

If you have a disability, such as a learning or physical disability, that would prevent you from taking a CLEP exam under standard conditions, you may request accommodations at your preferred test center. Contact your preferred test center well in advance of the test date to make the necessary arrangements and to find out its deadline for submission of documentation and approval of accommodations. Each test center sets its own guidelines in terms of deadlines for submission of documentation and approval of accommodations.

Accommodations that can be arranged directly with test centers include:
- ZoomText (screen magnification)
- Modifiable screen colors
- Use of a reader, amanuensis, or sign language interpreter
- Extended time
- Untimed rest breaks

If the above accommodations do not meet your needs, contact CLEP Services at clep@info.collegeboard.org for information about other accommodations.

Interpreting Your Scores

CLEP score requirements for awarding credit vary from institution to institution. The College Board, however, recommends that colleges refer to the standards set by the American Council on Education (ACE). All ACE recommendations are the result of careful and periodic review by evaluation teams made up of faculty who are subject-matter experts and technical experts in testing and measurement. To determine whether you are eligible for credit for your CLEP scores, you should refer to the policy of the college you will be attending. The policy will state the score that is required to earn credit at that institution. Many colleges award credit at the score levels recommended by ACE. However, some require scores that are higher or lower than these.

Your exam score will be printed for you at the test center immediately upon completion of the examination, unless you took College Composition. For this exam, you will receive your score four to six weeks after the exam date. Your CLEP exam scores are reported only to you, unless you ask to have them sent elsewhere. If you want your scores sent to a college, employer or certifying agency, you must select this option through My Account. This service is free only if you select your score recipient at the time you register to take your exam. A fee will be charged for each score recipient you select at a later date. Your scores are kept on file for 20 years. For a fee, you can request a transcript at a later date.

The pamphlet What Your CLEP Score Means, which you will receive with your exam score, gives detailed information about interpreting your scores. A copy of the pamphlet is in the appendix of this Guide. A brief explanation appears below.

How CLEP Scores Are Computed

In order to reach a total score on your exam, two calculations are performed.

First, your "raw score" is calculated. This is the number of questions you answer correctly. Your raw score is increased by one point for each question you answer correctly, and no points are gained or lost when you do not answer a question or answer it incorrectly.

Second, your raw score is converted into a "scaled score" by a statistical process called equating. Equating maintains the consistency of standards for test scores over time by adjusting for slight differences in difficulty between test forms. This ensures that your score does not depend on the specific test form you took or how well others did on the same form. Your raw score is converted to a scaled score that ranges from 20, the lowest, to 80, the highest. The final scaled score is the score that appears on your score report.

How Essays Are Scored

The College Board arranges for college English professors to score the essays written for the College Composition exam. These carefully selected college faculty members teach at two- and four-year institutions nationwide. The faculty members receive extensive training and thoroughly review the College Board scoring policies and procedures before grading the essays. Each essay is read and scored by two professors, the sum of the two scores for each essay is combined with the multiple-choice score, and the result is reported as a scaled score between 20 and 80. Although the format of the two sections is very different, both measure skills required for expository writing. Knowledge of formal grammar, sentence structure and organizational skills are necessary for the multiple-choice section, but the emphasis in the free-response section is on writing skills rather than grammar.

Optional essays for CLEP Composition Modular and the literature examinations are evaluated and scored by the colleges that require them, rather than by the College Board. If you take an optional essay, it will be sent to the institution you designate when you take the test. If you did not designate a score recipient institution when you took an optional essay, you may still select one as long as you notify CLEP within 18 months of taking the exam. Copies of essays are not held beyond 18 months or after they have been sent to an institution.

The CLEP Biology Exam

The Biology examination covers material that is usually taught in a one-year college general biology course. The subject matter tested covers the broad field of the biological sciences, organized into three major areas: molecular and cellular biology, organismal biology, and population biology.

The examination gives approximately equal weight to these three areas. The examination contains approximately 115 questions to be answered in 90 minutes. Some of these are pretest questions that will not be scored. Any time candidates spend on tutorials and providing personal information is in addition to the actual testing time.

Knowledge and Skills Required

Questions on the Biology examination require candidates to demonstrate one or more of the following abilities.

- Knowledge of facts, principles, and processes of biology
- Understanding the means by which information is collected, how it is interpreted, how one hypothesizes from available information, how one draws conclusions and makes further predictions
- Understanding that science is a human endeavor with social consequences

The percentages next to the main topics indicate the approximate percentage of exam questions on that topic.

33% Molecular and Cellular Biology

- Chemical composition of organisms
- Simple chemical reactions and bonds
- Properties of water
- Chemical structure of carbohydrates, lipids, proteins, nucleic acids
- Origin of life

Cells

- Structure and function of cell organelles
- Properties of cell membranes
- Comparison of prokaryotic and eukaryotic cells

Enzymes

- Enzyme-substrate complex
- Roles of coenzymes
- Inorganic cofactors
- Inhibition and regulation

Energy transformations

- Glycolysis, respiration, anaerobic pathways
- Photosynthesis

Cell division
- Structure of chromosomes
- Mitosis, meiosis, and cytokinesis in plants and animals

Chemical nature of the gene
- Watson-Crick model of nucleic acids
- DNA replication
- Mutations
- Control of protein synthesis: transcription, translation, posttranscriptional processing
- Structural and regulatory genes
- Transformation
- Viruses

34% Organismal Biology
- Structure and function in plants with emphasis on angiosperms
- Root, stem, leaf, flower, seed, fruit
- Water and mineral absorption and transport
- Food translocation and storage
- Plant reproduction and development
- Alternation of generations in ferns, conifers, and flowering plants
- Gamete formation and fertilization
- Growth and development: hormonal control
- Tropisms and photoperiodicity

Structure and function in animals with emphasis on vertebrates
- Major systems (e.g., digestive, gas exchange, skeletal, nervous, circulatory, excretory, immune)
- Homeostatic mechanisms
- Hormonal control in homeostasis and reproduction

Animal reproduction and development
- Gamete formation, fertilization
- Cleavage, gastrulation, germ layer formation, differentiation of organ systems
- Experimental analysis of vertebrate development
- Extraembryonic membranes of vertebrates
- Formation and function of the mammalian placenta
- Blood circulation in the human embryo

Principles of heredity
- Mendelian inheritance (dominance, segregation, independent assortment)
- Chromosomal basis of inheritance
- Linkage, including sex-linked
- Polygenic inheritance (height, skin color)

33% Population Biology

Principles of ecology
- Energy flow and productivity in ecosystems
- Biogeochemical cycles
- Population growth and regulation (natality, mortality, competition, migration, density, r- and K-selection)
- Community structure, growth, regulation (major biomes and succession)
- Habitat (biotic and abiotic factors)
- Concept of niche
- Island biogeography
- Evolutionary ecology (life history strategies, altruism, kin selection)

Principles of evolution
- History of evolutionary concepts
- Concepts of natural selection (differential reproduction, mutation, Hardy-Weinberg equilibrium, speciation, punctuated equilibrium)
- Adaptive radiation
- Major features of plant and animal evolution
- Concepts of homology and analogy
- Convergence, extinction, balanced polymorphism, genetic drift
- Classification of living organisms
- Evolutionary history of humans

Principles of behavior
- Stereotyped, learned social behavior
- Societies (insects, birds, primates)

Social biology
- Human population growth (age composition, birth and fertility rates, theory of demographic transition)

- Human intervention in the natural world (management of resources, environmental pollution)
- Biomedical progress (control of human reproduction, genetic engineering)

Chapter 2: How Science is Done

Scientific research serves two purposes:
1. To investigate and acquire knowledge that is theoretical *and*
2. To perform research that is of practical value

Science has the unique ability to serve humanity. Scientific research results from inquiry. An inquiring mind is thirsty, trying to find answers. An inquisitive person asks questions and wants to find answers. The two most important questions—why and how—are the starting points of all scientific inquiry.

The Scientific Method

Understanding how science is done can help you collect evidence and draw conclusions about scientific problems. In biology, a lot of evidence is gathered through observation. Paying attention to the living world can lead to some very interesting discoveries. However, there are times when collecting data can be more empirical. This means conducting an actual experiment to test a hypothesis.

Called the **scientific method**, this collection of steps is a universal way scientists test their ideas. The scientific method has eight steps:

1. State the problem
2. Collect background information
3. Establish a hypothesis
4. Perform the experiment
5. Analyze the data
6. Repeat the experiment
7. Draw conclusions
8. Report the results

Related to the scientific method is the term "scientific inquiry." While there are several definitions for this term, the overarching idea behind it is that all science starts with a question about the natural world. Things like, "How do sea turtles know on which beach to lay their eggs?" and "What feather coloration best attracts female blue jays?" are questions that can be tested using the scientific method. This is an important feature to remember. All science questions must be testable. Asking, "How much do you love your mom?" is not a testable question because the answer relies more on an intuitive perception than on empirical evidence.

Let us now examine each part of the scientific method to see how they all fit together.

State the Problem

Here, you want to identify some phenomenon that occurs in nature and ask a question about it. You might want to know how elephants select their mates or what the relationship is between acacia trees and certain ants. Most likely, this problem will come from some form of observation that has been made.

Collect Background Information

Once you have identified your problem, you will want to find out if anyone else has asked the same question. Now is the time to hit the library and the Internet to research what has already been done. You may find that many other people have already studied your problem. If that is the case, then you could either accept what they have already found *or* you could carry their research further. Let's say, for example, you wanted to know how long it takes chicken eggs to hatch in an incubator at 37°C. After doing your background research, you find several sources that say it takes 21 days at this temperature. A good scientist would ask himself how this knowledge could be expanded. Maybe you want to know how a change in temperature would change the outcome.

However you decide to change the experiment, you need to do more background research until you find something that has not been done before, but all research does not need to be original. It is perfectly fine to replicate previous investigations. If your results turn out the same way as those from the original study, great! This adds more support for its conclusions. If your results are different, that is perfectly OK, too. This then presents the question, "Why?"

Good science generates more questions than answers.

Establish a Hypothesis

Your hypothesis is what you want to study and what you think is going to happen during the experiment. The simplest, most-straightforward explanation is usually a good place to start.

Examples of good hypotheses include:

- If chicken eggs are kept at 35°C for 21 days, then the number of hatchlings will be reduced.
- If phosphorus and nitrogen are added to the soil, then the overall plant height will increase.
- If bacteria are exposed to periods of radiation, then the frequency of binary fission will increase.

Each statement here presents a conditional part and a determining part. Remember that the most important aspect of a hypothesis is that it must be testable.

The following are examples of bad hypotheses:

- I love chocolate chip cookies. (This is a personal opinion that's untestable.)
- Paramecia have cilia they use to move. (This is an observation, but it doesn't pose a testable hypothesis about the consequences of having cilia.)

Both of these statements are surely observations, but they are missing the testable part.

Perform the Experiment

Now the fun begins! This is the part of the scientific method where you will actually perform your experiment to test your hypothesis. It is important to create an experimental design that will allow you to collect the type of data you want. It is also important for every experiment to have a control, an independent variable, and a dependent variable.

The **control** in an experiment is the setup that remains unchanged. For example, in an investigation testing the impact of nitrogen on plant growth, a control setup would be a plant receiving no nitrogen. By having a setup like this, you can see if the nitrogen is the factor causing the plants to grow.

The **independent variable** is the factor that is being tested. In the aforementioned plant experiment, the nitrogen would be the independent

variable. If you also wanted to test other nutrients, those would also be part of the independent variable setup.

The **dependent variable** is the factor that changes in response to the independent variable. Again, in the plant experiment, you want to know how the addition of nitrogen or other nutrients impacts plant growth. The growth of the plants depends on the addition of the nutrients, so it is the dependent variable.

It is very important to only test one independent variable at a time. Think about what would happen if you tested nitrogen and phosphorus at the same time. How would you know which nutrient was causing the plants to grow?

Let's look at another example here. This time, fill out the information as you read through.

Here is the problem statement:

A student wants to investigate how long it takes a colony of bacteria to grow on an agar plate that contains one of three additives.

Here is the background:

It has been shown that certain nutrients when added to agar allow bacterial colonies to increase in size and grow at a more rapid rate.

What is a hypothesis that would test this problem?

Here is the experimental design:

The student took four agar plates and added nutrient 1 to the first, nutrient 2 to the second, and nutrient 3 to the third. The fourth plate was left alone. The plates were then inoculated with the bacterium and all placed into an incubator at 37°C for 24 hours. The student then counted the number of bacterial colonies on each plate.

What is the control setup in this experiment?

What is the independent variable?

What is the dependent variable?

Here is what you could have said for the hypothesis:

If nutrients are added to the agar plates, then the growth rates of the bacterial colonies will increase.

The control setup is the agar plate without any nutrients added.

The independent variable would be the addition of the nutrients to the agar plates.

The dependent variable is the growth of the bacterial colonies.

How did you do?

Analyze the Data

In this part of the scientific method, you look at the data you have collected to try to organize it. This is *not* the place where you try to figure out what it means. You just want to record it and put it in a form that best represents it. This may include a data table, a graph, a pie chart, or just listing different figures.

The type of graphic representation used to display observations depends on the type of data collected. **Line graphs** compare different sets of related data and help predict data. For example, a line graph could compare the rate of activity of different enzymes at varying temperatures. A **bar graph** or **histogram** compares different items and helps make comparisons based on the data. For example, a bar graph could compare the ages of children in a classroom. A pie chart is useful when organizing data as part of a whole. For example, a **pie chart** could display the percent of time students spend on various after-school activities.

As previously noted, the researcher controls the independent variable. When you create a graph, you place the independent variable on the x-axis (horizontal axis). The dependent variable is influenced by the independent variable and is placed on the y-axis (vertical axis). It is important to choose the appropriate units for labeling the axes. It is best to divide the largest value to be plotted by the number of blocks on the graph, and round to the nearest whole number.

It is likely that you will need to perform some kind of statistical analysis. Typical tests include median, mode, range, and Chi-square. Performing these tests can help to better understand what the numbers and values are that have been collected.

Repeat the Experiment

You will recall from earlier that all hypotheses need to be testable. Well, the experiments that are performed also need to be repeatable. This means that, using your experimental design, any researcher can replicate the experiment exactly the same way that you performed it originally. This is used to validate your results. If you do the experiment one time and discover that chickens hatch just as well at 25°C as they do at 37°C, but nobody can reproduce those results, your experiment will not have a lot of credibility.

The other reason to repeat the experiment is to increase the internal validity of the results. If a test is performed once, then it is highly possible the findings are a fluke of nature. However, if the experiment is performed hundreds of times (or as many as a plausible), and it comes out the same way most of the time, then there is a good possibility that the results are true. Remember this phrase, "the larger the sample size, the more accurate the results."

Draw Conclusions

Now that you have performed the experiment, analyzed the data, and repeated the experiment several times, it is time to figure out what your results mean. Assuming the results were the same, or close, each time you performed the experiment, it is safe to look for trends in the data. Which events happened most often, least often, and were outliers are factors that you should consider. The next thing to do is ask the question, "Why did the results appear as they did?" There is no right or wrong answer to this question, only good answers and bad ones. Your job as the scientist is to come up with possible explanations about why things happened the way they did and then support these claims with evidence.

In the bacterial growth experiment, let's say that colonies with nutrient 1 grew significantly faster than colonies on the other two experimental plates and the control plate. Why would this happen? Well, maybe nutrient 1 caused the genes of the bacteria to duplicate at a faster pace. Maybe the nutrient 1 mutated the bacteria's DNA in such a way that binary fission increased.

At this time, there is no way to know for certain. All you know is that the results of the plate with nutrient 1 added to it were different from those of the other plates. In your conclusion you would address this result, present ideas about why this may have happened, and then provide some further ideas for experimentation. Remember, science generates more questions than it does answers.

Report the Results

One of the most important aspects about the world of science is publishing one's findings. Remember back when you started thinking about what you wanted to investigate? What did you do to find background information? You looked into what had been published about the subject previously. Well, had all of these scientists not reported their findings, you would never have known what was already out there and gone ahead and wasted a lot of time replicating an experiment that had already been done.

There are several places where scientists publish their work. The main one is in a professional journal. A quick search for "academic biology journals" in a search engine brings up titles such as:

The New England Journal of Medicine
The Auk
Animal Behavior
Developmental Biology
The Journal of Cell Biology and Genetics
Marine Ecology Progress Series
Research Journal of Soil Biology
International Journal of Plant Biology and Research

The list goes on and on. These are just some of the printed journals that publish biology research. With the advent of the Internet, there are now thousands more online journals. You can also publish scientific research in books, magazine articles, short papers, and present findings at professional conferences. However it is done, it is arguably the most important step of the scientific method because it lets the rest of the world know what you are doing.

The Scientific Process

Science is limited by the available technology. For instance, it was only with the invention of the microscope that scientists like Robert Hooke and Antonie van Leeuwenhoek discovered cells, which were invisible to the naked eye. As technology improves, it allows more hypotheses to be tested in different ways, which, in turn, can lead to the development of theories and possibly laws. Data collection methods also limit scientific inquiry. Data may be interpreted differently on different occasions. The inherent limitations of scientific methodology produce results or explanations that are subject to change as new technologies emerge.

Hypothesis: An unproven idea or educated guess followed by research to best explain a phenomenon.

Theory: A statement of principles or relationships relating to a natural event or phenomenon that has been verified many times and accepted (for example, the Theory of Evolution, the Cell Theory).

Law: An explanation of events that occur with uniformity under the same conditions (for example, laws of nature, law of gravitation).

Scientific Facts

Facts are not always as finite as they appear. More commonly in science, information is a hypothesis or, once tested and confirmed, a theory. Theories exist for long periods and repeatedly receive challenges. Only when a theory has withstood every challenge and been proven to provide reproducible results does it become recognized as a law. It is the universal recognition that defines a theory as a scientific law.

Scientific Concepts

A concept is a general understanding or belief. Scientists challenge concepts. The purpose of the scientific method is to derive clear, unbiased data. Concepts, on the other hand, may be fraught with personal biases and gray areas, overly simplistic, or too encompassing. A scientist might examine a concept, and then try to confirm it by making and testing a hypothesis. In this way, scientific inquiry is more specific and concepts are more generalized.

Scientific Models

Models are the basis for greater understanding. Models are usually small-scale representations that help us understand a larger system. Models aid us by making unusually large or small items more concrete. Common models include the solar system and the DNA helix. It is important to note that models are created with information. How current and accurate the information is at the time of creation may make the model more or less useful later. For example, although Pluto has been considered a planet for many years, it is now considered a dwarf planet. This is due to the progressive nature of science; the more we learn, the more we are forced to reevaluate.

Biology Does Math Too!

Math, science, and technology share many common themes. All three use models, diagrams, and graphs to simplify a concept for analysis and interpretation. Patterns observed in these systems lead to predictions based on these observations. Another common theme among these three systems is equilibrium. **Equilibrium** is a state in which forces are balanced, resulting in stability. Static equilibrium is stability due to a lack of changes, and dynamic equilibrium is stability due to a balance between opposing forces.

Measurement and Notation Systems

Science uses the **metric system**, since it is accepted worldwide and allows the results of experiments, performed by different scientists around the world, to be compared to one another. The meter is the basic metric unit of length. One meter is 1.1 yards. The liter is the basic metric unit of volume. There are 3.846 liters to 1 gallon. The gram is the basic metric unit of mass. One thousand grams equals 2.2 pounds. The following prefixes define multiples of the basic metric units:

Prefix	Multiplying factor	Prefix	Multiplying factor
deca-	10X the base unit	deci-	1/10 the base unit
hecto-	100X	centi-	1/100
kilo-	1,000X	milli-	1/1,000
mega-	1,000,000X	micro-	1/1,000,000
giga-	1,000,000,000X	nano-	1/1,000,000,000
tera-	1,000,000,000,000X	pico-	1/1,000,000,000,000

The common instrument used for measuring volume is the graduated cylinder. The standard unit of measurement is milliliters (mL). To ensure accurate measurement, it is important to read the liquid in the cylinder at the bottom of the meniscus, the curved surface of the liquid.

The common instrument used in measuring mass is the triple beam balance. The triple beam balance can accurately measure tenths of a gram and can estimate hundredths of a gram. The ruler and meter stick are the most commonly used instruments for measuring length.

How to Manipulate Your Data

Data manipulation is important to experimental study. You will recall from the data analysis discussion above that the purpose of analysis is to look for trends that may appear and to tabulate them into a usable form. Data manipulation begins by altering one variable at a time, and then assessing the results. Are the results similar to the last time? What has changed? Has the situation improved or worsened? This process is part of the scientific method, where scientists make predictions and then experiment to test validity. Quite often, this process takes many alterations, and manipulating the data and experimental parameters is useful. We are fortunate to have technological advances to aid us in this area. Biologists use a variety of tools and technologies to perform tests, collect and display data, and analyze relationships. Examples of commonly used tools include computer-linked probes, spreadsheets, and graphing calculators.

Biologists use computer-linked probes to measure various environmental factors including temperature, dissolved oxygen, pH, ionic concentration, and pressure. The advantage of computer-linked probes, as compared to more traditional observational tools, is that the probes automatically gather data and present it in an accessible format. This property of computer-linked probes eliminates the need for constant human observation and manipulation.

Biologists use spreadsheets to organize, analyze, and display data. For example, conservation ecologists use spreadsheets to model population growth and development, apply sampling techniques, and create statistical distributions to analyze relationships. Spreadsheet use simplifies data collection and manipulation and allows the presentation of data in a logical and understandable format.

Graphing programs are another technology with many applications to biology. For example, biologists use algebraic functions to analyze growth, development and other natural processes. Graphing programs can manipulate algebraic data and create graphs for analysis and observation. In addition, biologists use the matrix function of graphing programs to model problems in genetics. The use of graphing programs simplifies the creation of graphical displays including histograms, scatter plots, and line graphs. Finally, biologists connect computer-linked probes, used to collect data, to graphing programs to ease the collection, transmission, and analysis of data.

While it is useful to manipulate data in discovery efforts, it is never acceptable to fabricate or falsely advertise your data.

Commonly Shared Scientific Ideals

Biological science is closely connected to other scientific disciplines and technology resulting in a tremendous impact on society and everyday life. Scientific discoveries often lead to technological advances. Conversely, technology is often necessary for scientific investigation and advances in technology often expand the reach of scientific discoveries. In addition, biology and the other scientific disciplines share several concepts and processes that help unify the study of science. Finally, because biology is the science of living systems, biology directly affects society and everyday life.

Science and technology, while distinct concepts, are closely related. Science attempts to investigate and explain the natural world, while technology attempts to solve human adaptation problems. Technology often results from the application of scientific discoveries, and advances in technology can increase the impact of scientific discoveries. For example, Watson and Crick used science to discover the structure of DNA and their discovery led to many biotechnological advances in the field of genomics. These technological advances greatly influenced the medical and pharmaceutical fields. The success of Watson and Crick's experiments, however, was dependent on the technology available. Without the necessary technology, the experiments would have been impossible or would have failed.

The combination of biology and technology has improved the human standard of living in many ways, but the negative impact of increasing human life expectancy and population on the environment is problematic. In addition, advances in biotechnology (for example, genetic engineering, cloning) produce ethical dilemmas that society must consider.

The following are the concepts and processes generally recognized as common to all scientific disciplines:

- Systems, order, and organization
- Evidence, models, and explanation
- Constancy, change, and measurement
- Evolution and equilibrium
- Form and function

Because the natural world is so complex, the study of science involves the **organization** of items into smaller groups based on interaction or interdependence. These groups are called **systems**. Examples of organization are the periodic table of elements and the three-domain classification scheme for living organisms. Examples of systems are the solar system, cardiovascular system, Newton's laws of force and motion, and the laws of conservation.

Order refers to the behavior and measurability of organisms and events in nature. The arrangement of planets in the solar system and the life cycle of bacterial cells are examples of order.

Scientists use **evidence** and **models** to form **explanations** of natural events. Models are miniaturized representations of a larger event or system. Evidence is anything that furnishes proof.

Constancy and **change** describe the observable properties of natural organisms and events. Scientists use different systems of **measurement** to observe change and constancy. For example, the freezing and melting point of a given substance and the speed of sound are constant under constant conditions. Growth, decay, and erosion are all examples of natural change.

Evolution is the process of change over a long period of time. While biological evolution is the most common example, one can also classify technological advancement, changes in the universe, and changes in the environment as evolution.

Equilibrium is the state of balance between opposing forces of change. Homeostasis and ecological balance are examples of equilibrium.

Form and **function** are properties of organisms and systems that are closely related. The function of an object usually dictates its form and the form of an object usually facilitates its function. For example, the form of the heart (for example, muscle and valves) allows it to perform its function of circulating blood through the body.

How the Scientific Method Works Outside of Science

We are constantly bombarded with information. There are always news reports describing a study of a new cancer drug, how climate change is impacting the world's oceans, or the relationship between saturated fats and high blood pressure. Many times the information presented on television or

in the newspaper is being reported by someone who does not know how to distinguish between science and pseudoscience, nor discriminate which facts may be true.

However one receives data, it is important to think about what has been presented. In the scientific realm, numbers are stronger than words. If you are presenting your own data, be sure to provide support to your claims by providing specific examples that back them up. If you are reading a newspaper story, look for the evidence that will make you believe it to be true. By using the scientific method, you will be more likely to catch mistakes, correct biases, and obtain accurate results. When assessing experimental data, use proper tools and mathematical concepts. Because people often attempt to use scientific evidence in support of political or personal agendas, the ability to evaluate the credibility of scientific claims is a necessary skill in today's society.

In evaluating scientific claims made in the media, public debates, and advertising, one should follow several guidelines. First, scientific, peer-reviewed journals are the most accepted source for information on scientific experiments and studies. One should carefully scrutinize any claim that does not reference peer-reviewed literature. Second, the media and those with an agenda to advance (for instance, advertisers and debaters) often overemphasize the certainty and importance of experimental results. One should question any scientific claim that sounds either too good to be true or overly certain. The media, especially commercials for new drugs, is riddled with the statement "scientifically proven" to show how great a product is. If you are aware of how science really works, you would know that science does not "prove" anything. It can suggest, lead us to believe, or support. To prove something means that every possible thing is known about it, without a shadow of a doubt. Does science work that way?

Chapter 3: Cellular Chemistry

Yes. You read that correctly. Chemistry. While this is a review book about the biology course you just took, it is important to know some basic chemistry in order to fully understand how biology works. Do not worry. This review will go over just enough for you to have a decent understanding of the relationship between the two fields. We'll save all the hard-core chemistry for when you take that class.

The Chemistry of Biology

Everything is made of atoms. An **atom** is the smallest particle of an element that exhibits the properties of that element. All of the atoms of a particular element are the same. The atoms of each element are different from the atoms of the other elements. An **element** is a substance that cannot be broken down into other substances. Today, scientists have identified 118 elements: 92 are found in nature and 26 are synthetic. Since elements are made of atoms, and all the atoms within an element are the same, these two terms are often used interchangeably.

All elements have a symbol. Most have one or two letters (some of the new synthetic elements have three). Some of the most important ones in biology include H_2, O_2, N_2, and C.

Atomic Structure

You are probably familiar with the overall structure of an atom. The most commonly accepted model is called the Bohr model and it contains a nucleus in the center and a cloud of rings floating around the outside (see figure 3.1).

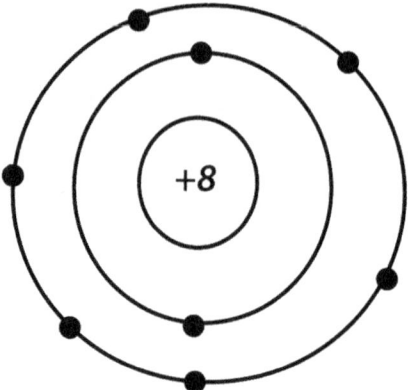

Figure 3.1. Bohr Model of Oxygen Atom

Inside the nucleus are the protons and neutrons. **Protons** are the positively charged subatomic particles of an atom. The number of protons in the nucleus is called the **atomic number** and identifies specifically which atom you are talking about. The **neutrons** are the neutral particles in the nucleus. They add to the mass of the atom and are important during radioactive decay. The collective mass of the protons and neutrons in the nucleus is called the **atomic mass**. Each element has a characteristic mass.

Outside of the nucleus is the **electron cloud**. Here, the **electrons** (the negatively charged subatomic particles) float around the nucleus in organized rings called energy shells. Under normal circumstances, the number of electrons equals the number of protons. However, since electrons are known to move, they are not an identifying characteristic of an atom.

Energy Shell	Maximum # of Electrons
1	2
2	8
3	16
4	32
5	50
6	72

Table 3.1. Electrons per Energy Shell

Each energy shell can hold a different number of electrons. The first shell is the one closest to the nucleus. Table 3.1 outlines the number of electrons each shell can hold.

It also important to note that each of the shells above the first one are composed of several smaller shells that hold varying numbers of electrons.

Compounds, Formulas, and Bonding – Oh My!

When elements join together they form a **compound**. The atoms within a compound are chemically combined, which results in each of the elements losing their individual identities and forming a new molecule with different properties.

Formulas are used to show the elements of a chemical compound. A **chemical formula** is a shorthand way of showing what is in a compound by using symbols and subscripts. The letter symbols tell which elements are involved and the number subscript tells how many atoms of each element are involved. No subscript is used if there is only one atom involved. For example, carbon dioxide is made up of one atom of carbon (C) and two atoms of oxygen (O_2), so the formula is CO_2. Examples of biologically important compounds include CO_2, $CaCO_3$, $C_6H_{12}O_6$, and H_2O.

Chemical bonds form when atoms with incomplete valence shells share or completely transfer their valence electrons. Valence electrons are those electrons found in the outer-most electron shell. For most elements, the outermost shell is not full, which is what allows them to bond with others. There are three types of chemical bonds: covalent bonds, ionic bonds, and hydrogen bonds.

Covalent bonding is the sharing of a pair of valence electrons by two atoms, for instance, between two hydrogen atoms. Each hydrogen atom has one valence electron in its outer shell; therefore the two hydrogen atoms come together to share their electrons (remember that the first electron shell can hold a maximum of two electrons). Some atoms share two pairs of valence electrons, like two oxygen atoms. This is a double covalent bond. Covalent bonds occur between elements that are non-metals.

How do you know if an atom is a metal or a non-metal? Well, on the periodic table there is often a highlighted boundary drawn that goes from number 5 (boron) down to number 84 (polonium) on the right hand side. The elements touching this boundary are called the metalloids because they share properties of both metals and non-metals. All of the elements to the left of the boundary

are metals. All of the elements to the right are non-metals. The only exception to this is hydrogen. Technically it is on the left of the boundary, so it would be a metal, but since hydrogen only has one electron, it needs to bond covalently.

Electronegativity describes the ability of an atom to attract electrons toward itself. The greater the electronegativity of an atom, the more it pulls the shared electrons toward itself. Electronegativity of the atoms determines whether the bond is polar or nonpolar. In **nonpolar covalent bonds**, the electrons are shared equally, thus the electronegativity of the two atoms is the same. This type of bonding usually occurs between two of the same atoms. A **polar covalent bond** forms when different atoms join together, as in hydrogen and oxygen to create water. In this case, oxygen is more electronegative than hydrogen so the oxygen pulls the hydrogen electrons toward itself.

Ionic bonds form when one electron is stripped away from its atom to join another atom. Ionic bonds occur between elements that are metals and that that are non-metals. For example, in the compound sodium chloride (NaCl), sodium is number 11, which means that it has 11 electrons. Remember that the first electron shell can hold 2 and the second can hold 8. That leaves just 1 electron in the outermost shell. Chlorine is atomic number 17. This means it has 17 electrons. It has 2 in the first shell, 8 in the second, and 7 in the third.

Because of these valence electrons, the sodium now has a +1 charge and the chloride now has a -1 charge. These charges attract each other to form an ionic bond. The "extra" electron of the sodium atom will be donated to the chlorine atom to fill its outmost shell. Ionic compounds are called salts. In a dry salt crystal, the bond is so strong it requires a great deal of strength to break it apart. However, if the salt crystal is placed in water, the bond will dissolve easily since the attraction between the two atoms decreases. Figure 3.2 shows the bonding. Each dot represents a valence electron.

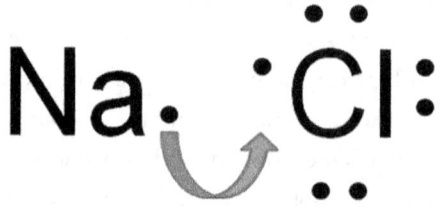

Figure 3.2. Ionic Bonding

The weakest of the three bonds is the **hydrogen bond**. A hydrogen bond forms when one electronegative atom shares a hydrogen atom with another electronegative atom. An example of a hydrogen bond occurs within a water molecule (H_2O) bonding with an ammonia molecule (NH_3). The H^+ atom of the water molecule attracts the negatively charged nitrogen in a weak bond. Weak hydrogen bonds are beneficial because they can briefly form, the atoms can respond to one another, and then break apart allowing formation of new bonds. Hydrogen bonding plays an important role in the chemistry of life.

pH and Buffers

The acidity of a solution is critical to living things. For example, the acidity of human blood is 7.4. Any variance plus or minus 0.01 points can cause death. The pH scale tells how acidic or basic a solution is. An acid is a solution that increases the hydrogen ion concentration of a solution, for example, hydrochloric acid (HCl). A base is a solution that reduces the hydrogen ion concentration by replacing it with another compound called hydronium, for example, in sodium hydroxide (NaOH).

The pH scale ranges from zero to 14 (see Figure 3.3). Seven is considered a neutral solution. Here, the amount of hydrogen equals the amount of hydronium. This is the pH of pure water. A pH of 6.9 and less is acidic. Stomach acid has a pH of 2.0. A pH of 7.1 and higher is basic. Common household bleach has a pH of 12.

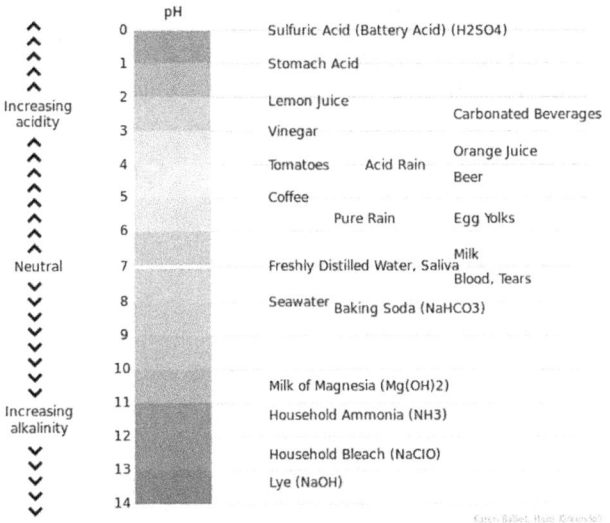

Figure 3.3. The pH Scale

A **buffer** accepts or donates H⁺ ions from or to the solution when they are in excess or depleted. Buffers are often used to neutralize acids in the environment and bring them back closer to 7. A common buffer is the antacid tablets some people take for heartburn. These tablets are basic in nature, so they are able to add hydronium ions to the acid. When hydrogen and hydronium interact, they form water, which as you know, is supposed to be neutral.

The pH of a substance has a dramatic effect on the environment as well. Acids greatly affect the biosphere. Acidic precipitation is rain, snow, or fog with a pH less than 5.6. Factories and cars spew acidic gases such as sulfur oxides and nitrogen oxides into the environment that react with water in the air to form acids that fall to Earth as precipitation. A change in pH in the environment can affect the solubility of minerals in the soil, which causes a delay in forest growth, and can also cause streams and ponds to become uninhabitable for fish and other wildlife.

Water. It's What Makes Life Possible.

Water is the most important compound on Earth. Without it, the planet would just be a large chunk of rock floating through space. Water has been called the "life blood" of the planet because it is what makes life as we know it possible.

Water is a polar covalent molecule composed of two hydrogen atoms and one oxygen atom (see Figure 3.4). The oxygen is the positive side and the hydrogen makes the negative side. This trait is important in the bonding of water molecules to each other. These hydrogen bonds are found in many biologicial molecules because they are easily broken and reassembled.

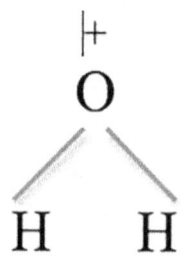

Figure 3.4. Water Molecule

In addition to only being found in its liquid form on Earth (solid water has been found on Mars), water has several other unique properties.

1. Water is called the **universal solvent**. The polarity of the water molecules allows many substances to dissolve within it. There is an old chemistry addage that says "like dissolves in like." Well, water both supports and refutes this claim because both covalent and ionic molecules will dissolve in it.
2. It takes a lot of energy to heat water. This is called its **heat of vaporization**. This trait is useful to living things because it allows for evaporative cooling. Many animals, including humans, remove excess heat from their bodies by having mechanisms that remove water from the body. When the water evaporates, it carries the heat away with it.
3. Water holds its heat for a long time. This is called its **specific heat**. Since water can hold its heat for a long period of time, it provides stable environments for those organisms living within it. The oceans and ponds stay warmer in the winter than the air does.
4. Water molecules like to stick to themselves and other things. This property is called **cohesion** (to each other) and **adhesion** (other things). Cohesive forces are useful because they create surface tension. This allows insects and other organisms to walk across the surface of water. Adhesive forces are important to plants. Water molecules stick to the sides of plants' internal tubes, which allows it to move up the stem from the roots.
5. Water's liquid form is more dense than its solid form. Ice floats on top of water. This provides a layer of insulation, which keeps more heat in.

The Molecules of Life

Molecules vary in size. Some, such as hydrogen gas are tiny. Others, like table sugar are quite large. Others, like many types of proteins, are enormous, each containing thousands or even millions of atoms. Regardless of its size, each molecule has a specific shape and structure. They are characterized by their chemical properties. These molecules often form together to form special compounds. There are four main chemical compounds found in the cells and bodies of living things. These include carbohydrates, lipids, proteins, and nucleic acids.

All of molecules of life are composed of smaller pieces. Each of these is called a **monomer**. The monomers join together to form **polymers**, which then fit together to make a large variety of molecules. The process of joining monomers into polymers is called a **condensation reaction** (also called dehydration synthesis). In this process, one molecule of water is removed between each of the adjoining molecules. In order to break apart the molecules in a polymer, water molecules are added between monomers, thus breaking the bonds between them. This process is called hydrolysis.

Functional Groups

All biological molecules are composed of carbon, hydrogen, and oxygen. These elements exist in varying proportions, depending upon the molecule. Many times the carbon forms a backbone of some kind to which the hydrogen and oxygen attach. At the same time, there are special compounds that also attach to these carbon skeletons that give them their particular jobs. These compounds are called **functional groups**. Table 3.2 outlines the most common functional groups and what roles they play.

Functional Group	Name	Properties
$-CH_3$	methyl	hydrophobic, component of DNA
$-OH$	hydroxyl	hydrophilic but no charge, forms alcohols
$-COOH$	carboxyl	hydrophilic, negative in solutions, found in lipids
$-NH_2$	amino	hydrophilic, positive in solutions, found in proteins
$-SH$	sulfhydryl	slightly hydrophobic, forms disulfide bonds, stabilizes protein structure
$-OPO_3^{-2}$	phosphate	hydrophilic, negatively charged, found in ATP
$-CH_3-OH$	carbonyl	hydrophilic, but no charge, found in sugars

Table 3.2. Functional Groups

Carbohydrates

Carbohydrates contain a ratio of two hydrogen atoms for each carbon and oxygen $(CH_2O)_n$. Carbohydrates include sugars and starches. They function to supply the organism with a quick release of energy. **Monosaccharides** are the monomers of carbohydrates and are the simplest sugars. The most common are glucose, fructose, and galactose. The chemical formula for all three of these molecules is $C_6H_{12}O_6$. They are called **isomers** because they have the same formula but a different shape (see Figure 3.5). These are the major nutrients for cells. During cellular respiration, the cells extract the energy from glucose molecules.

Figure 3.5. Isomers of Monosaccharides

Disaccharides are made by joining two monosaccharides by a condensation reaction. Maltose is the combination of two glucose molecules, lactose is the combination of glucose and galactose, and sucrose is the combination of glucose and fructose (see Figure 3.6). Notice that the formula for a disaccharide is twice that of a monosaccharide, $C_{12}H_{24}O_{11}$.

Figure 3.6. Disaccharide Structure

Polysaccharides consist of many monosaccharides joined together and may be structural or provide energy storage for the cell. Some of these molecules are huge! As energy stores, polysaccharides are hydrolyzed as needed to provide sugar for cells. Examples of polysaccharides include starch, glycogen, cellulose, and chitin.

Starch: a major energy storage molecule in plants. It is a polymer consisting of glucose monomers.

Glycogen: a major energy storage molecule in animals. It is made up of many glucose molecules.

Cellulose: found in plant cell walls, its function is structural. Many animals lack the enzymes necessary to hydrolyze cellulose, so it simply adds bulk (fiber) to the diet.

Chitin: found in the exoskeleton of arthropods and fungi. Chitin contains an amino sugar (glycoprotein).

Lipids

While carbohydrates are used for a quick release of energy, lipids are used by organisms for longer energy storage. Lipids also serve structural and endocrine functions. Lipids are composed of glycerol (an alcohol) and three fatty acids (see Figure 3.7). Lipids are hydrophobic (water fearing) and will not mix with water. There are three important families of lipids; fats, phospholipids and steroids.

Figure 3.7. Fatty Acids (left, vertical line) and Glycerol (right)

Fats consist of glycerol (alcohol) and three fatty acids. Fatty acids are long carbon skeletons with a carboxyl group at one end. The nonpolar carbon-hydrogen bonds in the tails of fatty acids are highly hydrophobic. At room temperature, fats are either solid or liquid. Solid fats, such as butter and lard, are called saturated and usually come from animals. They have single bonds between all of the carbon atoms in the carbon chain. Fats from plant sources

are called oils and are liquid at room temperature. They have at least one double bond in the carbon chain and have fewer hydrogen atoms attached.

Phospholipids are a vital component in cell membranes. In a phospholipid, one or two fatty acids are replaced by a phosphate group linked to a nitrogen group. These structures consist of a **polar** (charged) head that is hydrophilic (water loving) and a **nonpolar** (uncharged) tail, which is hydrophobic. This allows the membrane to orient itself with the polar heads facing the interstitial fluid found outside the cell and the nonpolar tails facing the internal fluid of the cell.

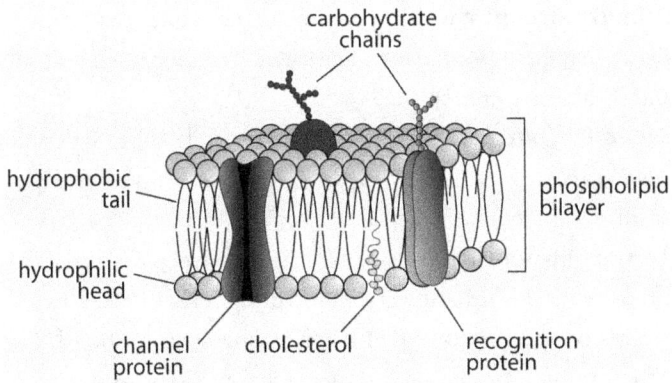

Figure 3.8. Lipid Bilayer of the Cell Membrane

Steroids are insoluble and are composed of a carbon skeleton consisting of four inter-connected rings. An important steroid is cholesterol, which is the precursor from which other steroids are synthesized. Hormones, including cortisone, testosterone, estrogen, and progesterone, are steroids. Their insolubility keeps them from dissolving in body fluids. While steroids are not technically lipids, they are insoluble in water, so they are classified in this group.

Proteins

Proteins comprise about fifty percent of the dry weight of animals and bacteria. Proteins function in structure and support (e.g., connective tissue, hair, feathers, and quills), storage of amino acids (e.g., albumin in eggs and casein in milk), transport of substances (e.g., hemoglobin), coordination of body activities (e.g., insulin), signal transduction (e.g., membrane receptor proteins), contraction (e.g., muscles, cilia, and flagella), body defense (e.g., antibodies), and as enzymes to speed up chemical reactions.

All proteins are made of 20 **amino acids**. Amino acids form through condensation reactions that remove water. The bond formed between two amino acids is called a **peptide bond**. An analogy can be drawn between the 20 amino acids and the alphabet. We can form millions of words using an alphabet of only 26 letters. Similarly, organisms can create many different proteins using the 20 amino acids, each of whose structure typically defines its function.

There are four levels of protein structure: primary, secondary, tertiary, and quaternary.

- **Primary structure** is the protein's unique sequence of amino acids. A slight change in primary structure can affect a protein's conformation and its ability to function.
- **Secondary structure** is the coils and folds of polypeptide chains, which are the result of hydrogen bonds along the polypeptide backbone.
- **Tertiary structure** results from bonds between the side chains of the amino acids. For example, disulfide bridges form when two sulfhydryl groups on the amino acids form a strong covalent bond. The tertiary structure of a protein is 3-dimensional in shape and it most directly influences how a protein is going to function. When proteins fall apart due to increased heat (called denaturing), it is because their tertiary structure has come undone.
- **Quaternary structure** is the overall structure of the protein from the aggregation of two or more polypeptide chains. For example, hemoglobin consists of four kinds of polypeptide chains.

Figure 3.9 shows the four different structural levels of proteins.

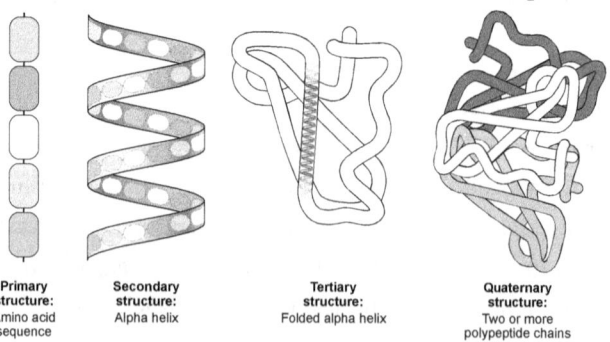

Figure 3.9. Protein Structure

Enzymes

Enzymes are large proteins that act as biological catalysts that speed up chemical reactions. They do this by lowering the **activation energy** for the reaction so it can get started sooner (*note – As you have learned, biology has many terms that can often be confusing and confused with others. The study of enzymes is one of the few places where this confusion is limited. If the word ends with the suffix –ase, it is an enzyme). They are not used up in a reaction and are recyclable. Each enzyme is specific for a single reaction. For example, the sugar sucrose is only broken down by the enzyme sucrase. Lipase would break down lipids. Amylase breaks down the sugar amylose. Easy Peasy!

Enzymes act on a substrate. The substrate is the material to be broken down or put back together.

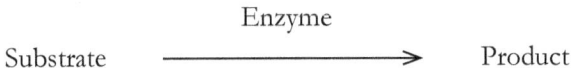

$$\text{Substrate} \xrightarrow{\text{Enzyme}} \text{Product}$$

The **active site** is the region of the enzyme that binds to the substrate. There are two ideas for how the active site functions. First, the **lock and key model** suggests that the shape of the enzyme is specific because it fits into the substrate like a key fits into a lock. The enzyme holds molecules close together so reactions can easily occur. The other idea is called the **induced fit model**. It states that an enzyme can stretch and bend to fit the substrate.

Factors Affecting Enzymes

Enzymes are very sensitive to changes in their environment. Even the smallest variations can cause the enzyme to stop working. Temperature and pH are probably the two factors that have the greatest impact on enzyme functionality. Enzymes can only work within their optimal ranges (see Figure 3.10). The optimal pH for enzymes is between 6 and 8, with a few enzymes whose optimal pH falls outside of this range, such as those found in the acidic stomach juices. The optimal temperature range is 30 – 40°C. If the temperature gets hotter than 40°C, the enzymes start to fall apart, or **denature.**

Something else to consider as a limiting factor of enzyme functionality is its concentration. Enzymes and substrates are limited in how much they can do at a given time. If those thresholds are reached, it does not matter how much more enzyme or substrate is added to the system. No more work can be done.

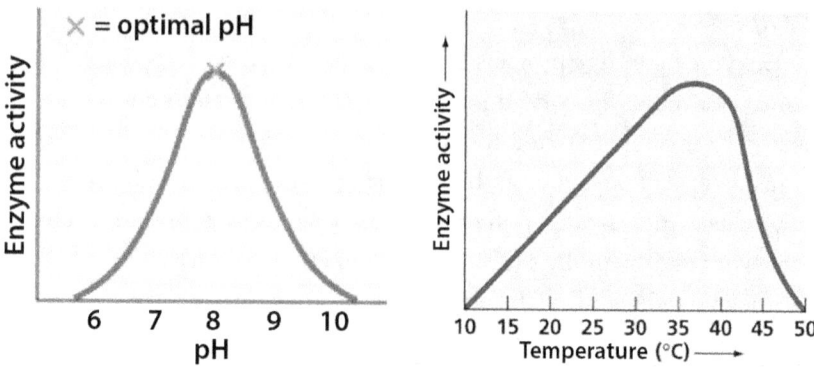

Figure 3.10. Optimal Enzyme Conditions

Cofactors and Coenzymes

Many times enzymes do not work alone. They often get the assistance of cofactors and coenzymes. **Cofactors** are inorganic minerals and **coenzymes** are organic vitamins. Coenzymes move from one enzyme to another, adding or taking away various chemicals from the substrate. A coenzymes binds to the active site, performs the reaction, and then separates in order to move to another substrate.

Nucleic Acids

Nucleic acids are arguably the most important of the biological molecules because they carry the instructions for every trait an organism possesses. The two main types of nucleic acids are DNA (deoxyribonucleic acid) and RNA (ribonucleic acid). The monomer of a nucleic acid is called a **nucleotide**. It consists of a 5-carbon sugar (deoxyribose in DNA, ribose in RNA), a phosphate group, and a nitrogenous base. There are five bases: adenine (A), thymine (T), cytosine (C), guanine (G), and uracil (U). Uracil is found only in RNA and replaces thymine. Table 3.3 provides a summary of nucleic acid structure.

	SUGAR	PHOSPHATE	BASES
DNA	Deoxyribose	Present	adenine, **thymine,** cytosine, guanine
RNA	Ribose	Present	adenine, **uracil,** cytosine, guanine

Table 3.3. Comparison of DNA and RNA

Adenine and guanine are called **purines**. They have a double carbon ring structure. Cytosine, thymine, and uracil are called **pyrimidines**. They have a single carbon ring. Because of the way these rings fit together, adenine will always pair with thymine, and guanine will always pair with cytosine. In RNA, uracil pairs with adenine. This is called **Chargaff's rule** or complementary base pairing. Hydrogen bonds hold the nitrogen bases together.

In 1953, James Watson, Francis Crick, and Maurice Wilkins discovered how the DNA molecule was structured. Using an x-ray photograph from another scientist named Rosalind Franklin, they figured out that the DNA molecule looks like a twisted ladder. The sides of the ladder are the sugar backbone and the rungs are the nitrogen bases. They called this structure a **double helix**. This allows for the symmetry of the DNA molecule seen below.

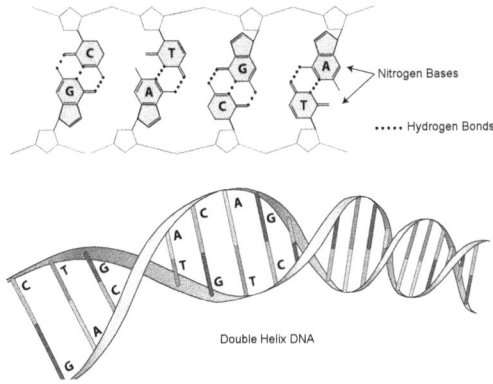

Figure 3.11. DNA Structure

Adenine and thymine (or uracil in RNA) are linked by two hydrogen bonds and cytosine and guanine are linked by three hydrogen bonds. Guanine and cytosine are harder to break apart than thymine (uracil) and adenine because of the greater number of bonds between the bases.

It is important to remember here that RNA is a single stranded molecule. During protein synthesis, it attaches to an exposed strand of DNA to get the needed code. There are three different types of RNA – mRNA, tRNA, rRNA, which will be discussed later.

DNA Replication

Each time the cells that make up an organism's body duplicate, all of their internal parts need to be copied. DNA is no different. It needs to copy

itself so that when new cells are formed, they each have the same genetic makeup as the parent cell. This process is called **DNA replication**.

DNA replication occurs during the interphase portion of the cell cycle. The process starts when the enzyme DNA helicase causes a segment of the DNA double helix to untwist into two separate strands. This breaks the hydrogen bonds holding them together and exposes the nitrogen bases on either side.

Floating around within the nucleus are free nucleotides. Once the double helix has unwound, the free nucleotides join up with the exposed bases. Remember, due to complementary base paring, A joins with T and C joins with G. An enzyme called DNA polymerase attaches the free nucleotides with those attached to the sugar backbone. Figure 3.12 shows the process of DNA replication.

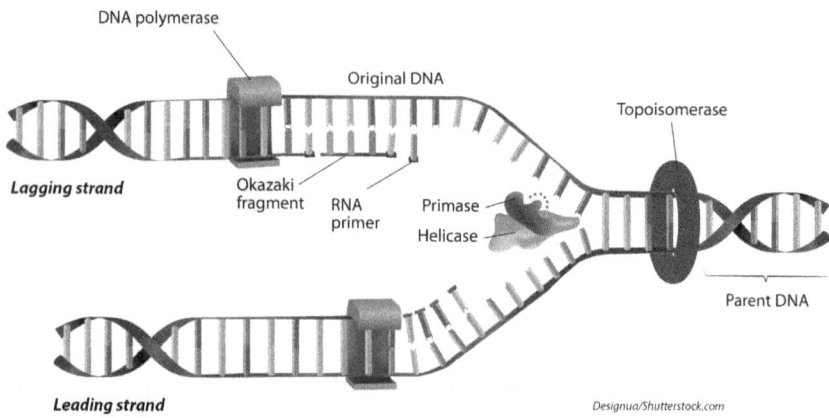

Figure 3.12. DNA Replication

When replication occurs, many times nucleotides from the ends of the strands get lost, which results in certain genes not being available to the organism anymore. To help protect chromosomes from such loss, chromosomes have structures called **telomeres** on their ends. These caps help keep the nucleotides from falling off. Telomeres are also believed to

play a role in aging. Human telomeres shorten with each DNA replication. Over time, they usually cease to function after 20–30 generations. However, scientists have recently been able to stop this death by exposing the telomeres to high levels of an enzyme called telomerase. These telomeres seem to live forever.

Did I Make a Mistake?

Errors happen all the time during DNA replication, for instance, when incorrect nitrogen bases get inserted or when strands align incorrectly. Fortunately, there are mechanisms in place to quickly repair them before any permanent damage can happen to the organism. Cells have three main methods of correcting errors.

- The first is simple proofreading. If DNA polymerase accidentally attaches a nucleotide that does not belong, the proofreading failsafe removes it.
- The second mechanism is called mismatch repair. Here, the DNA is scanned immediately after the free nucleotides are attached to remove any pairs that have been placed incorrectly.
- The third mechanism is called exclusion repair. This removes abnormal bases that have formed because of chemical damage and replaces them with new, functional ones.

Chapter 4: Energy of the Cell

One of the characteristics you need to be considered living is that you must use energy. Viruses are not considered alive because they do not have mechanisms in place to produce or metabolize energy. Cells, on the other hand, perform very complicated processes to create and use large amounts of energy.

Energy and Thermodynamics

The law of conservation of energy states that energy is neither created nor destroyed. Thus, energy changes form when energy transactions occur in nature. There are several types of energy, which you should be familiar with.

- **Thermal energy** is the total internal energy of objects created by the vibration and movement of atoms and molecules. Heat is the transfer of thermal energy.
- **Radiant energy** is the energy of electromagnetic waves. Light, visible and otherwise, is an example of radiant energy.
- **Electrical energy** is the movement of electrical charges in an electromagnetic field. Examples of electrical energy are electricity and lightning.
- **Chemical energy** is the energy stored in the chemical bonds of molecules. For example, the energy derived from gasoline is chemical energy.
- **Mechanical energy** is the potential and kinetic energy of a mechanical system. Rolling balls, car engines and body parts in motion exemplify mechanical energy.
- **Nuclear energy** is the energy present in the nucleus of atoms. Division, combination, or collision of nuclei release nuclear energy.

Because the total energy in the universe is constant, energy continually transitions between forms. For example, an engine burns gasoline converting the chemical energy of the gasoline into mechanical energy, a plant converts the radiant energy of the sun into the chemical energy found in glucose, or a battery converts chemical energy into electrical energy.

Heat energy is an example of relatively "useless" energy often generated during energy transformations. **Exothermic** reactions release heat and **endothermic** reactions require heat energy to proceed. For example, the human body is notoriously inefficient in converting chemical energy from food into mechanical energy. The digestion of food is exothermic and produces substantial heat energy.

Cells and Energy – The Two Big Processes

Cellular bioenergetics is the comparison of energy investment and the flow of energy through the cell. In this area of study, we ask if the product is worth the energy investment it requires. In the case of cells, we see that metabolism and reproduction are both worth the energy input because the cell profits in both areas. Photosynthesis results in usable energy, as does cellular respiration. It is important to note that optimal conditions will improve bioenergetics and can sometimes have an effect on the cellular pathway chosen.

Redox Reactions

Both photosynthesis and cellular respiration are called biochemical pathways. During these two events, energy molecules are shuttled from one place to another and the products of one are often used as the reactants of another. Many of the events that occur here are the result of something called an **oxidation-reduction reaction**.

When a molecule gets oxidized, it loses an electron. When a molecule gets reduced, it gains an electron. For example, during photosynthesis, NADPH is oxidized to NADP+ (the molecule loses an electron). When it gains that electron back, it becomes reduced.

There are a few pneumonic devices used to remember this process. One of them is

LEO the lion says GER

Here, LEO means Lose Electron—Oxidized and GER means Gain Electron—Reduced.

There are others, so whichever you remember is fine. However, it is highly recommended that you memorize one of them because there is good chance you will need to know how to do this.

Photosynthesis

You have been hearing about photosynthesis since you were in elementary school. You probably have it memorized as "the way plants make food from the Sun." Yes. This is correct. However, since you are now much older, wiser, and more sophisticated, you need to learn a more appropriate definition. **Photosynthesis** is the process by which autotrophs convert light energy into chemical energy.

You have seen this before. Here is the equation for photosynthesis:

$$CO_2 + H_2O + \text{energy (from sunlight)} \rightarrow C_6H_{12}O_6 \text{ (glucose)} + O$$

In words, carbon dioxide and water, with the input of energy from the Sun, are converted into a carbohydrate called glucose and oxygen. Keep this in mind as you continue to review. Visually, it can look like this (Figure 4.1):

Notice that we said "autotrophs" instead of "plants." This is because there are other organisms that perform this process. An **autotroph** (self-feeder) is an organism that makes its own food from the energy of the sun or other elements. Autotrophs include

- **photoautotrophs** that make food from light and carbon dioxide, releasing oxygen that can be used for respiration.
- **chemoautotrophs** that oxidize sulfur and ammonia. Some bacteria are chemoautotrophs.

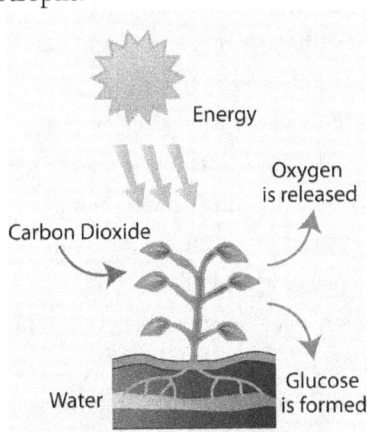

Figure 4.1. Visual Representation of Photosynthesis

The **chloroplast** is the site of photosynthesis in a plant cell (see Figure 4.2). Similar to mitochondria in a eukaryotic cell, chloroplasts contain a double membrane that increases its internal surface area. These membranes, called the **thylakoid** membranes, look like stacks of sandwich cookies piled on top of each other. A stack of thylakoids is called a granum (*pl.* grana). The thylakoids contain the green pigment **chlorophyll** that is used to capture light energy. All of the thylakoids are bathed in an ion-rich solution called **stroma**. Like the mitochondrial matrix, stroma contains ribosomes and DNA.

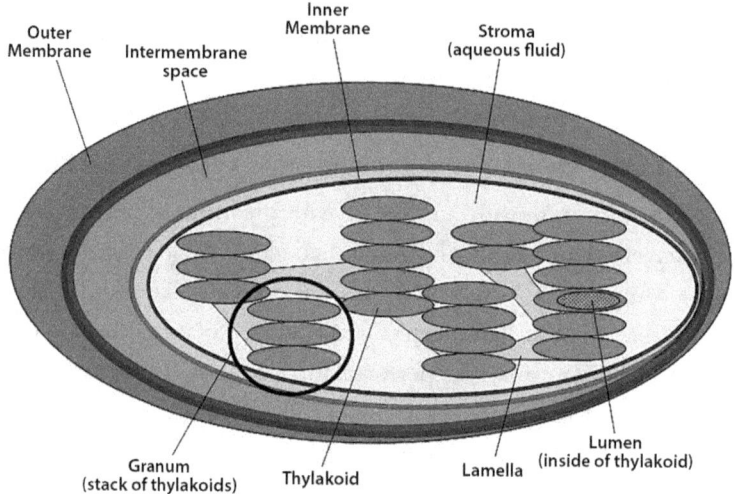

Figure 4.2. Structure of Chloroplast

Photosynthetic Pigments

Since light is the driving force behind photosynthesis, it is important to know something about how it behaves. Light can be reflected, absorbed, or transmitted. Remember that light from the Sun is considered white light. This means that it contains all of the wavelengths of the visible spectrum combined. When this white light hits the green leaves of a plant, most of the wavelengths get absorbed. Chlorophyll, being a green pigment, cannot absorb green light, so it gets reflected. This is why you see leaves as green. The main pigment is called chlorophyll *a*.

Plants also have several accessory pigments that help with photosynthesis. There are several different types of chlorophyll, all working to pass photons of light to chlorophyll *a*. Plants also have pigments called phycobilins and carotenoids that help. Phycobilins are those pigments that are red in color and reflect red wavelengths. Carotenoids are orange and yellow and reflect these

wavelengths. These are also the pigments you see in the fall when the leaves change color. Chlorophyll production decreases at this time of year, so the green coloration of the leaves goes away exposing the other colors present.

Visible light ranges in wavelengths of 750 nanometers (red light) to 380 nanometers (violet light). As the wavelength decreases, the amount of available energy increases. Light is carried as **photons**, which are fixed quantities of energy.

How It Happens

Photosynthesis occurs in two stages, the light-dependent reactions and the light-independent reactions (sometimes called the dark reactions, but this is a bit of a misnomer. These reactions can take place in the light. They just do not need it to proceed). The conversion of light energy to chemical energy occurs during light reactions. Here, chlorophyll absorbs light and uses the energy to split water, releasing oxygen as a waste product. The conversion of light energy to chemical energy is stored in the form of NADPH and ATP. You will remember that ATP (adenosine triphosphate) is the energy molecule of the cell. Both NADPH and ATP are then used in the Calvin cycle to produce sugar.

The high energy electrons are trapped by primary electron acceptors, which are located on the thylakoid membranes. These electron acceptors and the pigments form reaction centers called **photosystems** that are capable of capturing light energy. Photosystems contain a reaction-center chlorophyll that releases an electron to the primary electron acceptor. This transfer is the first step of the light reactions. There are two photosystems, named accordingly by their date of discovery, not their order of occurrence. Since photosystem II is at a lower energy than photosystem I, the electrons get accepted here first. They are then shuttled up to the higher energy photosystem I.

Photosystem I is composed of a pair of chlorophyll *a* molecules. Photosystem I is also called P700 because it absorbs light of 700 nanometers. Photosystem I makes ATP whose energy is needed to build glucose.

Photosystem II is also called P680 because it absorbs light of 680 nanometers. Photosystem II produces ATP + $NADPH_2$ and the waste gas oxygen.

The production of ATP is termed **photophosphorylation** due to the use of light. Photosystem I uses cyclic photophosphorylation because the pathway occurs in a cycle. It can also use noncyclical photophosphorylation,

which starts with light and ends with glucose. Photosystem II uses noncyclical photophosphorylation only.

So, at the end of the light-dependent reactions:

- Light has been absorbed.
- Photosystem II has shuttled energy to photosystem I.
- Water molecules have been split.
- Energy in the form of ATP and NADH has been produced.

The second stage of photosynthesis is the **Calvin cycle.** Here, the main job is to produce as many carbon molecules as possible. This occurs in the stroma of the chloroplast. If you recall the photosynthesis equation you saw earlier, you know that the only source of carbon to make sugars comes from carbon dioxide. Carbon dioxide enters the leaves through small openings on their undersides called **stomata** (these are also the same openings that release the oxygen produced, as shown in Figure 4.3).

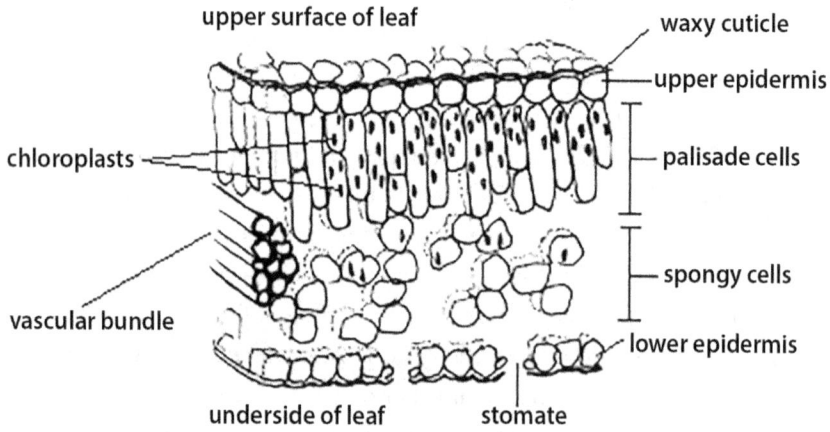

Figure 4.3. Cross section of Leaf

The Calvin cycle requires a lot of energy. The NADPH and ATP produced during the light reactions are used as the power needed for changing carbon to carbohydrates. Being a cycle, the energy, shuttle molecules, and enzymes are used over an over again. The major events of the Calvin cycle are summarized below.

1. A molecule of CO_2 reacts with a 5-carbon molecule called ribulose bisphosphate (RuBP).

2. This 6-carbon molecule splits into two molecules of phosphoglycerate (PGA).
3. ATP and NADPH convert PGA to glyceraldehyde-3-phosphate (G3P). This is 3-carbon sugar, also sometimes called PGAL (phosphoglyceraldehyde).
4. G3P then drops off a carbon while the rest is recycled into more RuBP.

It takes six turns of the Calvin cycle to produce one molecule of glucose (do the math: every turn drops off one carbon atom. Glucose needs six carbons).

Variations on a Theme

The normal production of glucose happens through what is known as the C3 pathway. The C3 refers to the 3-carbon molecule called PGAL that is produced during the Calvin cycle. There are times, however, when normal photosynthesis does not work. Certain environmental conditions have caused the evolution of alternate pathways to accomplish photosynthesis.

During the C4 pathway, carbon dioxide gets fixed at lower concentrations. C4 plants, as they are known, fix the carbon into a 4-carbon molecule called **oxaloacetate.** The carbon is first converted into a 3-carbon molecule, which then gets converted to the 4-carbon molecule. C4 plants are able to keep their stomata closed more often, which reduces water loss.

The other alternative pathway for fixing carbon is called the CAM pathway. CAM plants live in very dry conditions so their stomata are always closed during the day to prevent water loss. They can only fix carbon at night. These plants have special enzymes that work to convert the oxaloacetate into the needed sugar molecules.

There are many events that happen during the two stages of photosynthesis. It is very easy to get them confused with each other and with the different steps of cellular respiration (as you will see below). To help sort things out, Figure 4.4 summarizes the two steps of photosynthesis.

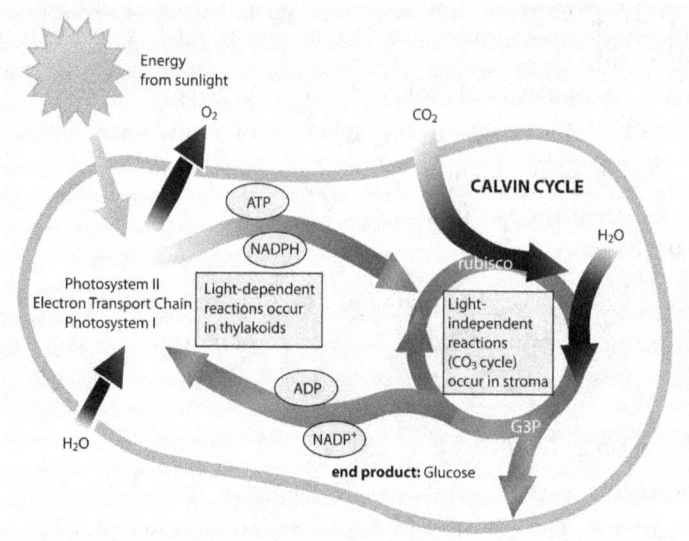

Figure 4.4. Photosynthesis Steps Compared

Cellular Respiration

Figure 4.5 summarizes the relationship between photosynthesis and cellular respiration, which we will cover in this section.

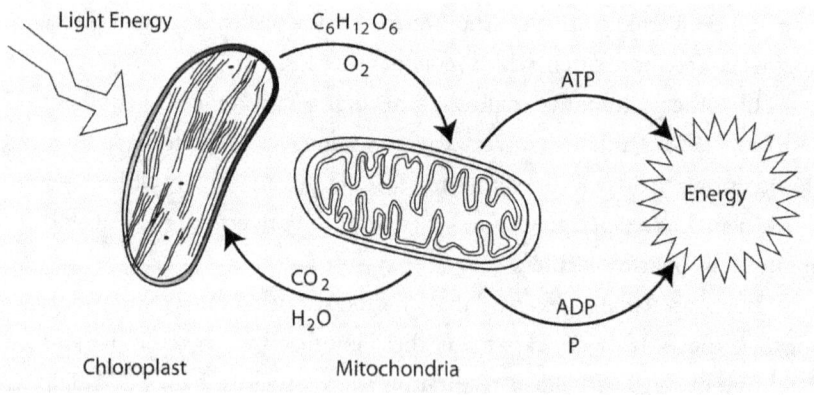

Figure 4.5. Relationship of Photosynthesis and Cellular Respiration

Cellular respiration is the metabolic pathway by which food (e.g., glucose) is broken down to produce energy in the form of ATP. Both plants and animals use cellular respiration to create energy for metabolism. As in photosynthesis, cellular respiration requires a series of oxidation-reduction reactions that release energy when electrons are transferred between molecules.

The biochemical pathway that occurs during cellular respiration can be summarized in the following equation:

$$C_6H_{12}O_6 + 6O_2 \rightarrow 6CO_2 + 6H_2O + \text{energy (ATP)}$$

To describe what is happening in words you would say that carbohydrates (in this case glucose) and oxygen are converted into carbon dioxide, water, and energy in the form of ATP. You will remember that ATP is the energy molecule of the cell. It is formed through a process called **phosphorylation**. Here, adenosine diphosphate (ADP) is added to a single inorganic phosphate. This forms ATP (adenosine triphosphate). The ATP is then broken apart to release the energy it contains. Phosphorylation takes energy, so that poses the question, Why bother putting ATP together if it is just going to be taken apart again? Well, it turns out that the amount of energy to add a phosphate to ADT is far less than is produced when ATP is broken apart, so there is a significant net gain in energy.

Glycolysis

The first step of cellular respiration is called **glycolysis**, which occurs in the cytoplasm of the cell and does not require oxygen. Overall, it is a very complicated process that involves the change of a glucose molecule into a compound called pyruvate. A specific enzyme catalyzes each intermediate step.

As glucose is converted, there is an initial input of energy, but an overall net gain. Glucose (which has six carbons) is broken into two 3-carbon molecules. For each of these molecules, a molecule of NADH (the cell's other energy carrier) is produced. Additionally, two molecules of ATP are generated for each of the 3-carbon sugars, but two of them are used up during the conversions. This leaves a net gain of two ATP.

Even though glycolysis produces ATP, it is only two percent of the entire amount that will be created from the breakdown of the glucose molecule.

Kreb's Cycle

Beginning with pyruvate, which was the end product of glycolysis, the following steps occur before entering the **Krebs cycle.**

1. Pyruvic acid is changed to acetyl-CoA (coenzyme A). This is a 3-carbon pyruvic acid molecule that has lost one molecule of carbon dioxide (CO2) to become a 2-carbon acetyl group. Pyruvic acid loses a hydrogen to NAD+, which is reduced to NADH.

2. Acetyl CoA enters the Krebs cycle. For each molecule of glucose it started with, two molecules of Acetyl CoA enter the Krebs cycle (one for each molecule of pyruvic acid formed in glycolysis).

The **Krebs cycle** (also known as the citric acid cycle) takes place within the mitochondria and requires oxygen (Figure 4.6). Note that, like chloroplasts, mitochondria have a double membrane system. The inner membrane forms a series of folds called **cristae** that are used to increase the internal surface area. The inner membrane forms a compartment called the **matrix**, which is where the Krebs cycle takes place.

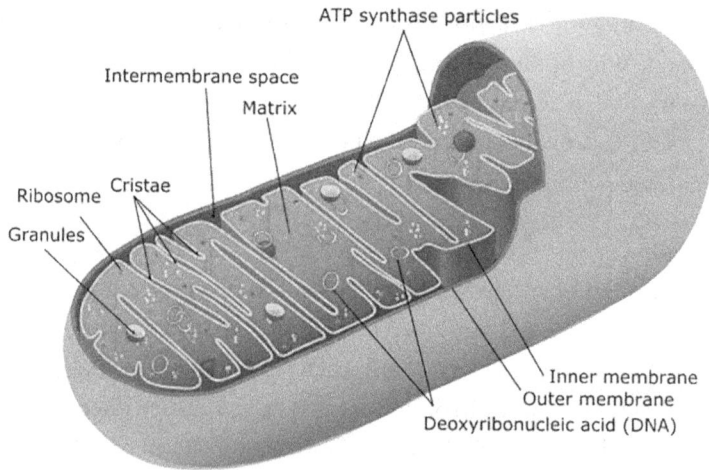

Figure 4.6. Internal Structure of Mitochondrion

There are four major steps involved in the Krebs cycle. First, the 2-carbon acetyl CoA combines with a 4-carbon molecule to form a 6-carbon molecule of citric acid. Next, two carbons are lost as carbon dioxide (CO_2) and a 4-carbon molecule is formed, which is available to join with CoA to form citric acid again (hence the "cycle"). Since we started with two molecules of CoA, two turns of the Krebs cycle are necessary to process the original molecule of glucose. In the third step, eight hydrogen atoms are released and picked up by FAD and NAD (vitamins and electron carriers).

Each turn of the Krebs cycle produces on1e molecule of ATP, 1 molecule of FADH, and 3 molecules of NADH, and carbon dioxide as a waste product. Since there were 2 molecules of pyruvate created at the end of glycolysis, these numbers are doubled.

This completes the breakdown of glucose. At this point, cellular respiration has created a total of 4 molecules of ATP, 2 from glycolysis and 1 from each of the two turns of the Krebs cycle. A total of 6 molecules of carbon dioxide have been released, 2 prior to entering the Krebs cycle and 2 for each of the two turns of the Krebs cycle. In addition, 12 carrier molecules have been made, 10 NADH and 2 $FADH_2$. These carrier molecules will carry electrons into the electron transport chain.

Notice that the Krebs cycle in itself does not produce many ATP molecules. Instead, it functions mostly in the transfer of electrons that are subsequently used in the electron transport chain to generate large numbers of ATP molecules.

Electron Transport Chain

In the **Electron Transport Chain** (ETC), NADH transfers electrons from glycolysis and the Krebs cycle to the first molecule in the chain of molecules embedded in the inner membrane of the mitochondrion.

Most of the molecules in the electron transport chain are proteins. Nonprotein molecules are also part of the chain and are essential for the catalytic functions of certain enzymes. The electron transport chain does not make ATP directly. Instead, it breaks up a large free energy drop into a more manageable one. The chain uses electrons to pump H^+ ions across the mitochondrial membrane (remember all the folds of the cristae?). The H^+ gradient is used to form ATP synthesis in a process called **chemiosmosis** (oxidative phosphorylation). ATP synthetase and energy, generated by the movement of hydrogen ions coming from NADH and $FADH_2$, builds ATP from ADP on the inner membrane of the mitochondria. Each NADH yields three molecules of ATP (10 x 3) and each $FADH_2$ yields two molecules of ATP (2 x 2). Thus, the electron transport chain and oxidative phosphorylation produces 34 ATP and the net gain from the whole process of respiration is 36 molecules of ATP.

Table 4.1 compares the input and output of energy during the different steps of cellular respiration.

Process	# ATP produced (+)	# ATP consumed (-)	Net # ATP
Glycolysis	4	2	+2
Acetyl CoA	0	2	-2
Krebs cycle	1 per cycle (2 cycles)	0	+2
Electron transport chain	34	0	+34
Total			+36

Aerobic versus Anaerobic Respiration

You know that glycolysis can generate ATP in the presence of oxygen (aerobic) or in its absence (anaerobic). So far we've discussed aerobic respiration. The Krebs cycle needs oxygen in order to progress, but sometimes oxygen is not available. Organisms living in anaerobic environments perform a process called **fermentation**. During fermentation, ATP is generated by substrate level phosphorylation if enough NAD^+ is present to accept electrons during oxidation. In anaerobic respiration, NAD^+ is regenerated by transferring electrons to pyruvate. There are two common types of fermentation.

Alcoholic fermentation involves the conversion of pyruvate into ethanol. To do this, carbon dioxide is released from the pyruvate. This form of fermentation is important to the baking and adult beverage industries. When baking bread, yeast eat the sugar that is added to the mixture, converting it to alcohol and releasing carbon dioxide. This is what causes the bread to rise.

Have you ever been running and developed a cramp in your side? If so, then you have experienced lactic acid fermentation. **Lactic acid fermentation** involves the reduction of pyruvate by NADH to form lactate as a waste product. Animal cells and some bacteria that do not use oxygen use lactic acid fermentation to make ATP. Lactic acid forms when pyruvic acid accepts hydrogen from NADH. Unlike alcoholic fermentation, which is unidirectional, lactic acid fermentation is reversible. When oxygen returns to the system, lactic acid dissipates and normal aerobic respiration continues.

Energy remains stored in lactic acid or alcohol until it is needed. This is not an efficient type of respiration. When oxygen is present, aerobic respiration occurs after glycolysis.

Both aerobic and anaerobic pathways oxidize glucose to pyruvate through the process of glycolysis and both pathways employ NAD^+ as an oxidizing agent. A substantial difference between the two pathways is that in fermentation, an organic molecule such as pyruvate or acetaldehyde is the final electron acceptor. In respiration, the final electron acceptor is oxygen. Another key difference is that respiration yields much more energy from a sugar molecule than fermentation does. Respiration can produce up to 18 times more ATP than fermentation.

One final comment here to help you organize your thinking about cellular respiration. You will notice that throughout this section, the process has been called "cellular respiration" not "respiration." This is because "respiration" is another term for breathing. Air enters the lungs and carbon dioxide is expelled. While there is definitely a connection between these two processes, they are very different. Respiration happens at the organismal level and results in an exchange of gases. Cellular respiration happens at the cellular level and results in the production of energy.

Chapter 5: Cell Structure and Function

All living things are composed of cells. In fact, that is the number one requirement needed to be considered alive. The first person to look at cells was Dutch lens grinder Anton van Leeuwenhoek in the 1600s. By manipulating pieces of glass of different thicknesses, he was able to focus light so that an image would be magnified. In essence, he invented the first microscope. Leeuwenhoek's work was carried on by English scientist Robert Hooke. He examined pieces of cork under the microscope and observed that the cork looked like little rooms that were similar to those used by monks in a monastery. These rooms were called "cells," so Hooke took the term for what he observed under the microscope.

In the 1800s, three scientists, each working independently, studied the behavior and anatomy of cells and came up with what today is known as the cell theory. It states that

- all living things are composed of one or more cells;
- cells are the basic units of structure and function in an organism; and
- all cells arise only from other cells.

Cell Diversity

Cells range in size from smaller than 10 micrometers to the size of a large cantaloupe (the ostrich egg is the largest single cell. It weighs in at over 3 pounds!). All cells have a cell membrane that encloses all of the internal parts and regulates what enters and leaves. They also all have some form of genetic material. This material regulates the cell's activities and is also used for passing on traits to the next generation.

Prokaryotic Cells

The two main types of cells are called prokaryotic and eukaryotic. **Prokaryotic cells** lack a true membrane-bound nucleus and other internal structures. This cell type is found in the organisms known as archaea and bacteria. The archaea are separate from the bacteria because of where they live and how they metabolize energy. They live in extreme environments, such as hot springs, inside of volcanoes, and in extremely salty environments. They are anaerobic (live without oxygen) and are believed to be the precursors of life on Earth.

The bacteria (shown in Figure 5.1) have no defined nucleus or nuclear membrane. The DNA, RNA, and ribosomes float freely within the cell. The cytoplasm has a single chromosome condensed to form a **nucleoid.** This nucleoid contains all of the information needed for the cell to function. Pieces of it also get transferred to other bacterial cells during **conjugation**. Bacterial cells have a thick cell wall made up of amino sugars (glycoproteins) that provides protection, gives the cell shape, and keeps the cell from bursting. The antibiotic penicillin targets the **cell wall** of bacteria. Penicillin works by disrupting the cell wall, thus killing the cell.

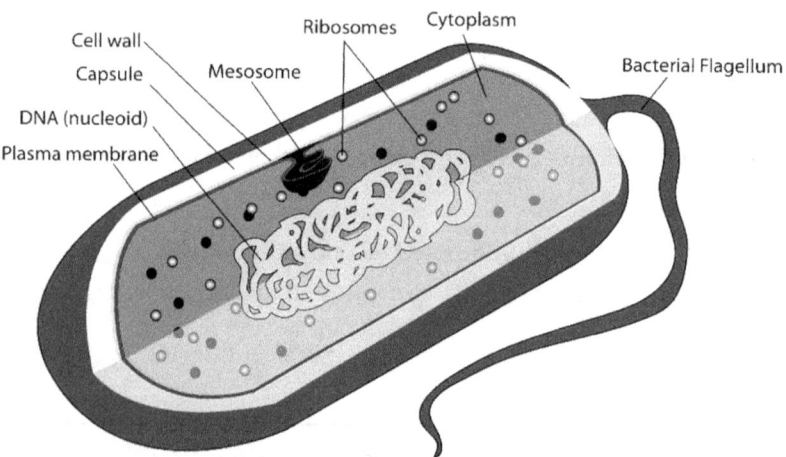

Figure 5.1. Typical Bacterial Cell

The cell wall surrounds the **cell membrane** (plasma membrane). The cell membrane consists of a lipid bilayer that controls the passage of molecules in and out of the cell. Some prokaryotes have a capsule, made of polysaccharides, surrounding the cell wall for extra protection from higher organisms.

Many bacterial cells have appendages used for movement called **flagella**. Some cells also have pili, which are a protein strand used for attachment. Pili are also used for sexual conjugation (where bacterial cells exchange DNA).

Prokaryotes are the most numerous and widespread organisms on earth. There are more bacteria present than anything else. Picture a field of grass. There are thousands of bacteria living on each blade.

Many people are scared of bacteria. What they fail to realize is that most bacteria are completely harmless to people. In fact, many are actually beneficial. Bacteria are used to make medicines that prevent disease, produce food, and help with digestion. Bacteria will be discussed in more detail later on.

Eukaryotic Cells

Eukaryotic cells are found in protists, fungi, plants, and animals. Most eukaryotic cells are larger than prokaryotic cells. They contain many **organelles**, which are membrane-bound areas with specific functions. Their cytoplasm contains a cytoskeleton, which provides a protein framework for the cell. The cytoplasm also supports the organelles and contains the ions and molecules necessary for cell function. The cytoplasm is contained by the plasma membrane. The plasma membrane allows molecules to pass in and out of the cell and can also bend inward to engulf outside material.

Figure 5.2 shows a generalized animal cell.

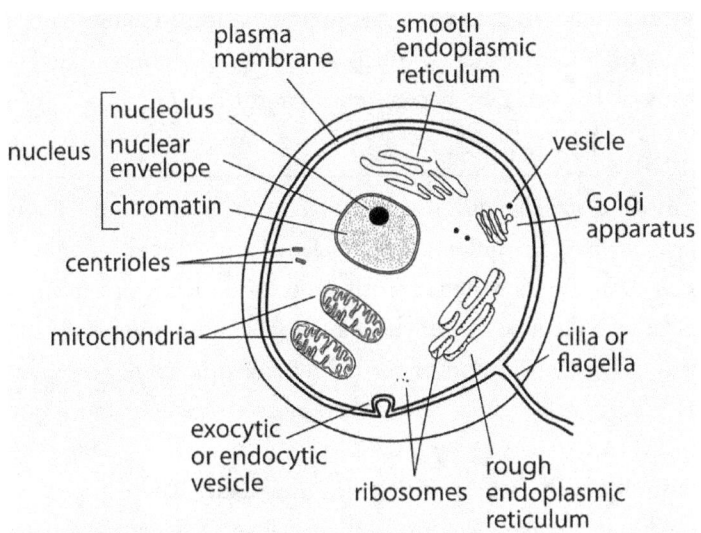

Figure 5.2. Animal Cell

Table 5.1 compares the differences between prokaryotic and eukaryotic cells.

Eukaryotic Cells	Prokaryotic Cells
between 10-100 microns	smaller than 10 microns
large ribosomes	small ribosomes
no plasmids	plasmids present
linear chromosomes	circular chromosomes
aerobic respiration	aerobic or anaerobic respiration
cytoskeleton present	cytoskeleton absent
plant and some protist cells have cell wall	all have cell wall
can be unicellular or multicellular	all are unicellular
membrane-bound nucleus and organelles are present	membrane-bound organelles and nucleus are lacking

The **endosymbiotic theory** of the origin of eukaryotes states that eukaryotes arose from symbiotic groups of prokaryotic cells. According to this theory, smaller prokaryotes lived within larger prokaryotic cells, eventually evolving into chloroplasts and mitochondria. Chloroplasts are the descendants of photosynthetic prokaryotes, and mitochondria are likely the descendants of bacteria that were aerobic heterotrophs.

Organelles

The most significant difference between prokaryotes and eukaryotes is that eukaryotes have a **nucleus**. The nucleus is the "brain" of the cell and contains all of the cell's genetic information in the form of chromosomes. These chromosomes consist of chromatin, which are complexes of DNA and proteins. The chromosomes are tightly coiled to conserve space while providing a large surface area. The nucleus is also the site of transcription of DNA into RNA.

The **nucleolus** is where ribosomes are made. There is at least one of these dark-staining bodies inside the nucleus of most eukaryotes. The nuclear envelope consists of two membranes separated by a narrow space. The envelope contains many pores that let RNA out of the nucleus.

Ribosomes are the site for protein synthesis. Ribosomes may be free floating in the cytoplasm or attached to the endoplasmic reticulum. There may be up to a half a million ribosomes in a cell, depending on how much protein the cell makes.

The **endoplasmic reticulum** (ER) is folded and has a large surface area. It is the transport system of the cell and allows for transport of materials through and out of the cell. Smooth endoplasmic reticulum does not contain ribosomes on the surface and is the site of lipid synthesis. Rough endoplasmic reticulum has ribosomes on its surface and aids in the synthesis of proteins that are membrane-bound or destined for release to other parts of the organism.

Many of the products made in the ER proceed to the Golgi apparatus. The **Golgi apparatus** functions to sort, modify, and package molecules that are made in other parts of the cell (like the ER). These molecules are sent either out of the cell or to other organelles within the cell. The Golgi apparatus looks like a stack of pancakes. When materials leave the Golgi apparatus to be sent outside of the cell, a portion of them fuse with the cell membrane, forming a vesicle. This vesicle then opens to the external environment to release the material. As such, the Golgi apparatus must be located close to the cell membrane for it to function properly.

Lysosomes are found mainly in animal cells. These contain digestive enzymes that break down food, unnecessary substances, viruses, damaged cell components, and, eventually, the cell itself. It is believed that lysosomes play a role in the aging process.

Mitochondria are large organelles that are the site of cellular respiration, the process of ATP production that supplies energy to the cell. Muscle cells have many mitochondria because they use a great deal of energy. Mitochondria have their own DNA, RNA, and ribosomes and are capable of reproducing by binary fission if there is a great demand for additional energy. Mitochondria have two membranes: a smooth outer membrane and a folded inner membrane (see Figure 5.3). The folds inside the mitochondria, called **cristae,** provide a large surface area for cellular respiration to occur.

Plastids are found only in photosynthetic organisms. They are similar to mitochondria in that they both have a double membrane structure and also have their own DNA, RNA, and ribosomes. They can reproduce if there is a need for increased capture of sunlight.

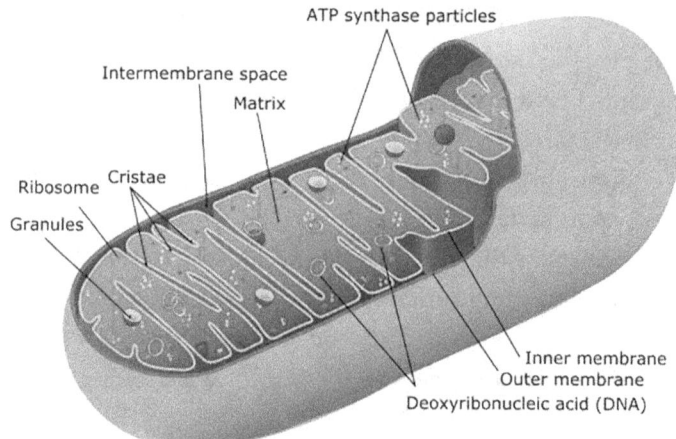

Figure 5.3. Mitochondrion

There are several types of plastids. **Chloroplasts** are the site of photosynthesis. The stroma is the chloroplast's inner membrane space. The stroma encloses sacs called **thylakoids** that contain the photosynthetic pigment chlorophyll. The chlorophyll traps sunlight inside the thylakoid to generate ATP, which is used in the stroma to produce carbohydrates and other products (see Figure 5.4). The **chromoplasts** make and store yellow and orange pigments. They provide color to leaves, flowers, and fruits. The **amyloplasts** store starch and are used as a food reserve. They are abundant in roots like potatoes.

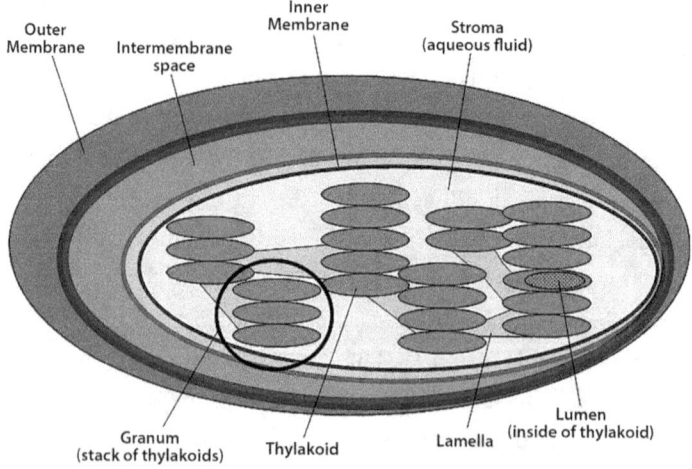

Figure 5.4. Chloroplast cross-section

Found only in plant cells, the **cell wall** is composed of cellulose and fibers. It is thick enough for support and protection, yet porous enough to allow water and dissolved substances to enter. **Vacuoles** are found mostly in plant cells. They hold stored food, water, and pigments. Their large size allows them to fill with water in order to provide **turgor pressure**. When the vacuole fills with water, the cell membrane is forced against the cell wall, which adds rigidity to the cell. Conversely, when water is lost from the vacuole, the cell membrane pulls away from the cell wall. Lack of turgor pressure causes a plant to wilt.

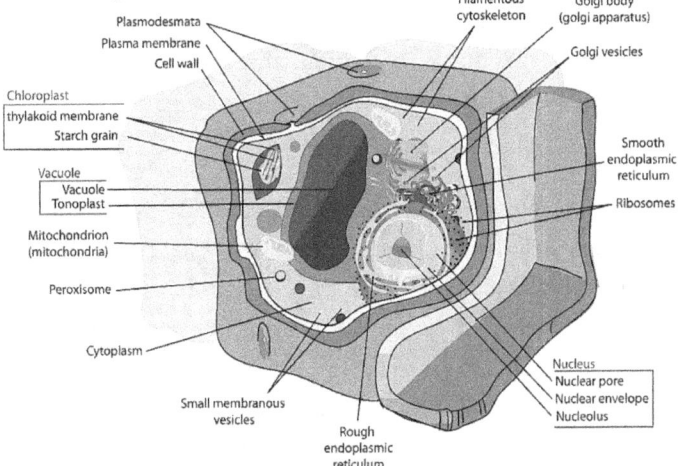

Figure 5.5. Plant Cell

The **cytoskeleton**, found in both animal and plant cells, is composed of protein filaments attached to the plasma membrane and organelles. The cytoskeleton provides a framework for the cell and aids in cell movement. Three types of fibers make up the cytoskeleton:

1. **Microtubules:** The largest of the three fibers, they make up cilia and flagella for locomotion, for instance in sperm cells, cilia that line the fallopian tubes, and tracheal cilia. Centrioles also are composed of microtubules. They aid in cell division by forming the spindle fibers that pull the cell apart into two new cells. Centrioles are not found in the cells of higher plants.
2. **Intermediate filament:** Intermediate in size, they are smaller than microtubules, but larger than microfilaments. They help the cell keep its shape.

3. **Microfilaments:** Smallest of the three fibers, they are made of the proteins actin and myosin (as in muscle tissue). They function in cell movement like cytoplasmic streaming, endocytosis, and amoeboid movement. It is the microfilaments that pinch the two cells apart after cell division, forming two new cells.

Getting Into and Out of the Cell

Since a cell is the smallest unit to be considered living, it makes sense that it has quite a complicated metabolism. All of the organelles you just reviewed produce materials (proteins, waste produces, etc.) that need to leave the cell or enter other ones. It is the cell membrane that regulates what, where, when, and how materials can enter or leave.

Cell Membrane

You know that the cell membrane is composed of two layers of lipids. This makes it waterproof and able to survive quite well being bathed by extracellular fluids (imagine what would happen if the cell membrane were composed of carbohydrates. Wicked Witch of the West, anyone??) Located within these two lipid layers are a variety of proteins and cholesterol that are floating around, which creates the characteristic **fluid mosaic** look of the membrane.

The cell membrane is called **selectively permeable.** This means that it only allows certain things to pass through it. If particles are too larger or have the incorrect charge, they cannot always enter (there are exceptions to this that will be described later). Found on the surface of the cell membrane are many specialized proteins that serve as a way for cells to recognize others. For example, egg cells can only be fertilized by sperm of the same species (usually—we are not talking about hybrids here). The proteins on the surface of the egg cell prevent sperm from other species from entering it and passing on its genes. Also, once the egg has been fertilized, these proteins instantly send a signal across the entire membrane to prevent additional sperm from fertilizing it.

Located within the cell membrane are additional proteins called channel proteins. These embedded structures are what allow small particles to pass through the membrane. Channel proteins have an electric charge to them. If the particle trying to get through them has the same charge, it will be repelled. The particle will then have to travel around the cell until it finds a channel with the opposite charge.

Water and oxygen can enter the cell whenever necessary. Oxygen is able to pass through the membrane at will, but water needs a little bit of help. A water molecule is too large to pass directly through the membrane, so it gets in and out through tiny pores called aquaporins. These openings do not ever prevent water from coming or going, but provide the way it gets in an out.

Figure 5.6 shows a cross-section of the cell membrane. Missing are the aquaporins.

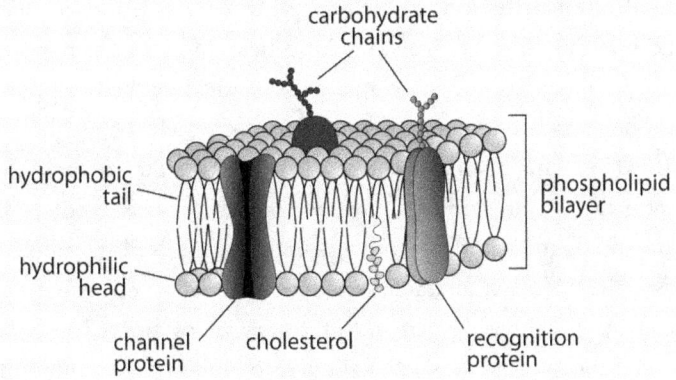

Figure 5.6. Cross-section of Cell Membrane

Put Your Left Foot In. Take Your Left Foot Out

OK. So we have established that the cell membrane is the crossing guard for all materials entering and exiting the cell. But how exactly does this happen? Sure, water moves through aquaporins or oxygen moves directly, but how does the cell "know" when it is time from more things to come in or leave?

The answer lies around a **concentration gradient**. With a concentration gradient, molecules move from areas of high concentration to areas of low concentration. Picture a beaker filled with water. A student then puts in a few drops of blue food coloring. The coloring initially globs together, but after a little time, the entire beaker has become blue. The water is called the **solvent** (does the dissolving), and the food coloring is called the **solute** (what gets dissolved). When referring to situations like this, it is usually the solute that moves. In this example, there is a lot of food coloring in that one spot, so to even itself out, its molecules will move all around the beaker until they are evenly distributed. The movement of the food coloring molecules is called diffusion. The equalization of those molecules throughout the beaker is called **equilibrium**.

Cell Transport

Diffusion is the movement of molecules without energy from an area of high concentration to an area of low concentration. In the motion described above, the food coloring had a very high concentration compared to the beaker of water. Its molecules moved down their gradient until they were evenly spread out. Diffusion does not take energy. As long as there is a high concentration, the molecules will always move down the gradient.

In **facilitated diffusion**, molecules still move down their concentration gradients, but this time they need a little help. When certain particles are trying to enter the cell, the channel proteins already described help move them through. The particle enters the channel proteins, which change their shape to accommodate it. They then push the particle through the opening. This form of diffusion also requires no energy, just a concentration gradient.

A specific form of diffusion that is important to cells is called osmosis. **Osmosis** is the diffusion of water through a selectively permeable membrane (remember this?). Water is essential to the survival of the cell, so it has to be able to come and go as necessary. If there is too much water within the cell, it will leave through the aquaporins to the extracellular environment. If the cell is dehydrated, then water from the extracellular environment enters the cell (picture watering a plant that has wilted. The water enters the roots, travels up the stem, and enters the cells of the leaves).

Osmotic Potential

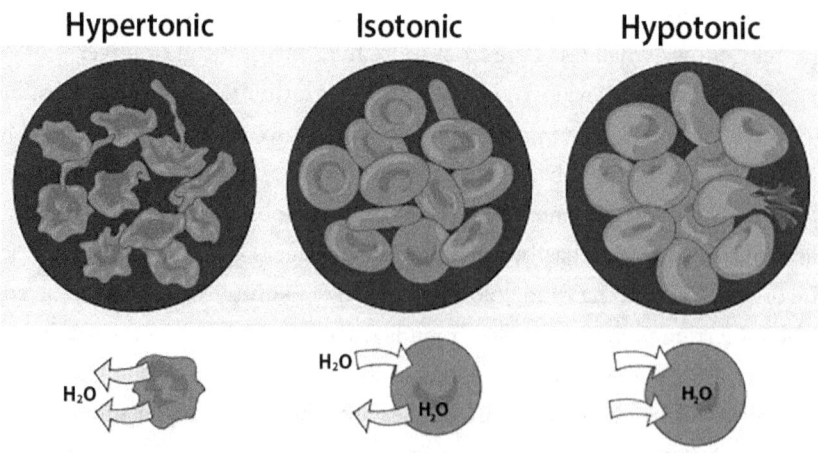

Figure 5.7. Osmotic Potential

Figure 5.7 shows what happens to blood cells when they are placed into different environments. In a **hypertonic** environment, there is more solute outside of the cell than there is inside, so water will leave the cell in a process called **plasmolysis** that causes the cell to shrink.

In an **isotonic** environment, the solute concentration is equal inside and outside of the cell. Water still moves, but the amount entering equals the amount leaving.

A **hypotonic** environment has more solute within the cell than in the external environment. Since the water needs to flow down its concentration gradient, it flows into the cell, causing it to swell. This swelling is what creates turgor pressure inside of plant cells.

Active Transport

Sometimes molecules need to move *up* their concentration gradients, swimming against the current as it were. As you realize, this effort takes energy (in the form of ATP). There are many examples of active transport in biology, but one in particular needs mentioning here. The sodium-potassium pump is a mechanism used to transmit electrical signals across a neuron using two atoms are called ions, meaning they each have a charge. They establish concentration gradients across the membrane of the neuron. When they receive a signal from the brain, more of each molecule is forced across membrane, against their gradients. This allows the signal to move.

There are other forms of active transport in the cell. These involve the taking in or expulsion of large particles (things are too large to diffuse across the cell membrane).

Endocytosis is the taking in of large particles. During this process, the particle (say a large carbohydrate) approaches the cell membrane. The membrane then wraps itself around the particle until it is full enclosed within a vesicle. The vesicle then detaches from the cell membrane and travels to the lysosomes, where its digestive enzymes dissolve the vesicle to get the carbohydrate. This is also the common way an amoeba feeds.

Endocytosis can be divided into two specific parts. **Pinocytosis** is the taking in of dissolved molecules—sort of like "cell drinking." **Phagocytosis** is the taking in of large food particles.

Exocytosis is the opposite of endocytosis. The proteins or waste products within the cell are moved towards the cell membrane (usually by the Golgi apparatus). The cell membrane then folds in, around the particles,

enclosing them entirely within a vesicle. Since the cell membrane cannot just open up to release the particles, the vesicle fuses with the cell membrane. With the continuity of the membrane intact, the outer part of the vesicle opens to the extracellular environment, releasing the particles.

How Cells Divide: Mitosis and Meiosis

The purpose of cell division is to provide growth and repair of body (somatic) cells and to replenish or create sex cells for reproduction. The two main forms of cell division are called mitosis and meiosis. **Mitosis** is the division of somatic cells, and **meiosis** is the division of sex cells (eggs and sperm).

Before we start discussing how cells divide, we must first consider the structure of a chromosome. Remember that a chromosome is nothing more than DNA wrapped around special proteins called **histones**. Each half of a chromosome is called a chromatid and is held together by a structure called a centromere. Figure 5.8 shows what a chromosome looks like.

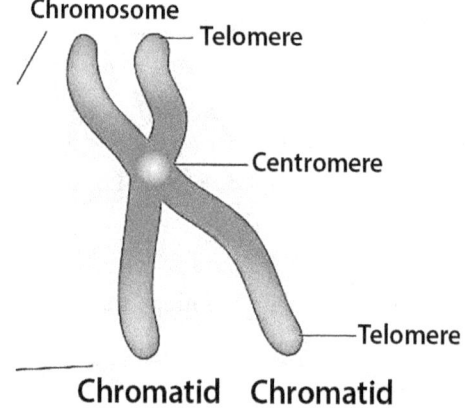

Figure 5.8. Chromosome Structure

Mitosis

There are two main parts to mitosis, the **mitotic (M) phase** and **interphase.** Depending on how your teacher taught this to you, you may or may not recognize the individual steps outlined here. Since the cell cycle is continuous, some people do not identify the individual events that occur. For the purposes here, we will use the steps of mitosis since they help identify the movements happening within the cell.

In the mitotic phase, mitosis and cytokinesis divide the nucleus and cytoplasm, respectively. The mitotic phase is the shortest phase of the cell cycle. Interphase is the stage where the cell grows and copies the chromosomes in preparation for the mitotic phase. Depending on the type of cell involved, this phase can last between a few hours and several days.

During interphase, there are three stages of growth. During the **G_1** (gap) period, the cell grows and metabolizes. It enters this phase immediately after completing the previous mitotic phase. Next, the cell enters the **S** (synthesis) period, when it copies all of its DNA. It is at this time when errors can occur. After the DNA is duplicated, the cell enters the **G_2** (gap) period, when the cell makes new proteins and organelles in preparation for cell division. Between each of these steps there is a checkpoint phase. Here, the cell monitors what has happened and corrects any errors that have occurred. If there is a problem at one of the checkpoints, there is the possibility that the rest of interphase will not progress.

As previously mentioned, the mitotic phase is a continuum of change, but we can divide the events that take place can into five distinct stages: prophase, prometaphase, metaphase, anaphase, and telophase.

During **prophase**, the cell goes through the following steps continuously, without stopping. First, the chromatin condenses to become visible chromosomes. Next, the nucleolus disappears and the nuclear membrane breaks apart, freeing the chromosomes for the movement they are about to undergo. The mitotic spindles, composed of microtubules, form that will eventually pull the chromosomes apart. Finally, the cytoskeleton breaks down, and the centrioles push the spindles to the poles or opposite ends of the cell.

During **prometaphase**, the nuclear membrane fragments and allows the spindle microtubules to interact with the chromosomes. Each chromosome attaches to the spindle by its centromere.

Metaphase begins when the chromosomes slide along the spindle fibers and align themselves along the equator of the cell. This phase is often the easiest to identify through a microscope.

During **anaphase**, the centromeres split in half and homologous pairs separate. The chromosomes are pulled to the poles at opposite ends of the cell.

The last stage of mitosis is called **telophase**. Here, two nuclei form, each with a full set of DNA that is identical to the parent cell. The nucleoli

become visible and the nuclear membrane reassembles. The spindle fibers disappear and the chromosomes unwind back into DNA.

Now that all of the genetic information has been duplicated, it is time for the cell to actually split into two. This process, called **cytokinesis** is slightly different between animal cells and plant cells. In an animal cell, the cell membrane pinches off in the center and ultimately pulls apart. In plant cells, however, the cell wall keeps the cell membrane from completely separating. Instead, a structure called a cell plate will form between the two new nuclei. The cell membrane then realigns itself along this cell plate to form two new cells.

The two resulting cells are called daughter cells and are genetically identical to the parent cell. Once cytokinesis occurs, each of the new daughter cells immediately enters back into the G_1 phase of interphase to start the process all over again.

Figure 5.9 outlines the steps of mitosis in an animal cell.

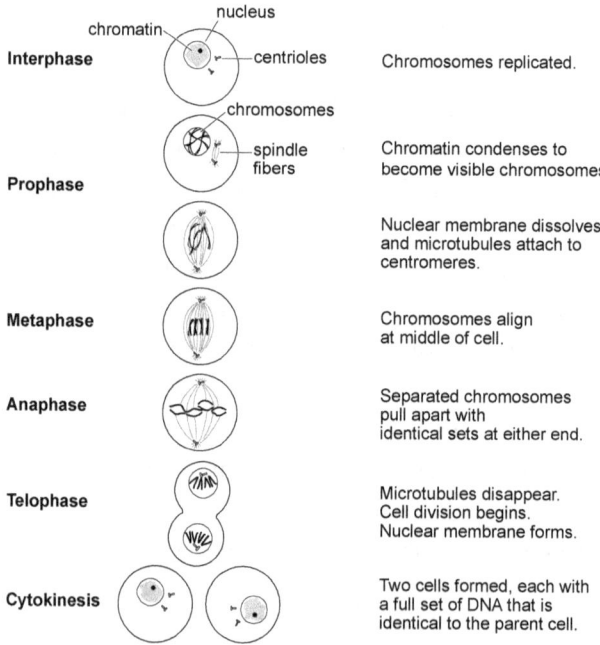

Figure 5.9. Animal Cell Mitosis

Meiosis

Meiosis is similar to mitosis, but instead there are two cell divisions instead of just one to produce cells that have half the chromosomes of the parent cell. In order to reduce the chromosome number by one-half it is necessary for each cell to divide twice. This way, when the sperm and egg join during fertilization, the correct diploid number is reached.

Similar to mitosis, meiosis is preceded by interphase during which the chromosomes replicate. The steps of meiosis are a bit more complicated than those of mitosis because it is producing cells that have half the chromosome number and are genetically diverse. The steps of meiosis are as follows:

1. **Prophase I** – The replicated chromosomes condense and pair with homologues in a process called **synapsis**. This forms a **tetrad**. Once the tetrad is formed a small piece of each pair trades places. This is called crossing over and serves to increase the genetic diversity of the offspring. This is why you do not look exactly like your parents. Also, as in prophase of mitosis, the nucleus dissolves, the spindle fibers form, and the DNA condenses into chromosomes.
2. **Metaphase I** – The homologous pairs of chromosomes attach to the spindle fibers and then slide to line up along the equator of the cell.
3. **Anaphase I** – The spindle fibers pull one chromosome from each tetrad towards the poles of the cell.
4. **Telophase I** – The homologous chromosome pairs gather into nuclei. Each pole now has a haploid chromosome set. Telophase I occurs simultaneously with cytokinesis.

There is no interphase between telophase I and prophase II.

5. **Prophase II** – A spindle apparatus forms, the chromosomes condense, and the nuclear membrane dissolves.
6. **Metaphase II** – Sister chromatids line up in equator of cell. The centromeres divide and the sister chromatids begin to separate.
7. **Anaphase II** – The separated chromosomes move to opposite ends of the cell. Because of crossing over and independent assortment, each new cell will have a different genetic makeup.
8. **Telophase II** – The chromosomes gather again into nuclei. Cytokinesis occurs, resulting in four haploid daughter cells.

Spermatogenesis (sperm production) and **oogenesis** (egg production) are the same in terms of the steps involved and the production of haploid

gametes. However, spermatogenesis produces four haploid cells, each containing the same number of chromosomes. Oogenesis produces one haploid cell that contains all of the genetic material and three "dead" cells called polar bodies. These **polar bodies** are reabsorbed by the female's body.

Think about this in terms of the reproductive life of an organism (let's discuss humans). Human males start being able to reproduce somewhere around 13 years of age. They then produce sperm throughout their entire lives. Yes, even 85 year old men are able to produce children. However, human females have a finite reproductive life. They start egg production about the same time as males, but sometime around age 50 they stop (through menopause). Considering human females release, on average, one egg per month (minus any time she may be pregnant), the total number of egg cells produced can be calculated.

Figure 5.10 details the steps of meiosis.

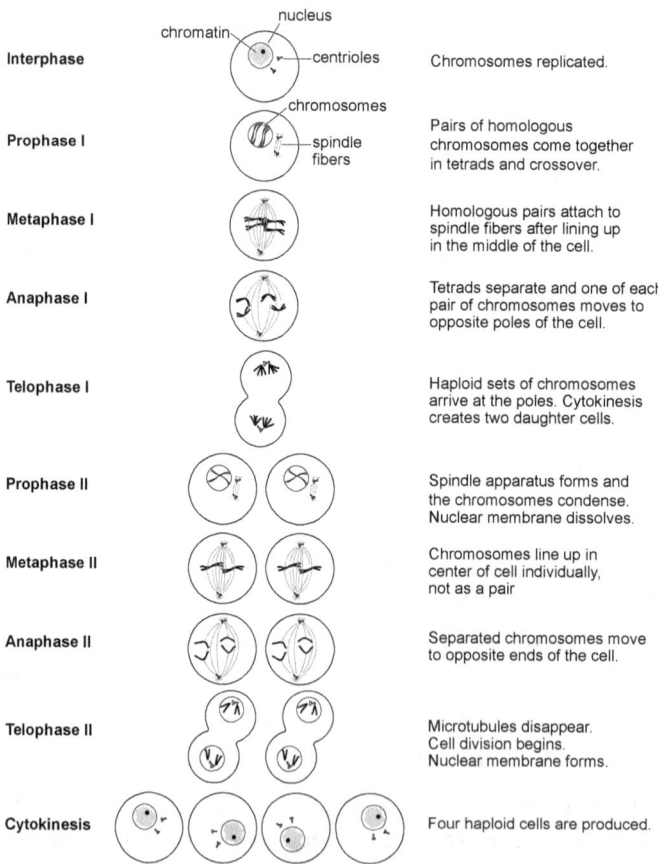

Figure 5.10. Meiosis

Chapter 6: Genetics

Gregor Mendel was an Austrian monk who is recognized as the father of genetics. His work in the late 1800s is the basis of all current knowledge of heredity. While he had no knowledge of DNA or genes, Mendel realized there were factors (now known as **genes**) that were transferred from parents to their offspring. Mendel worked with pea plants and fertilized the plants himself, keeping track of subsequent generations, which led to his laws of genetics. Mendel is also recognized as one of the first scientists to quantify his work. This means he performed thousands of trials in order to validate his findings.

Mendel also introduced the idea of probability into his work. **Probability** is the chance that something will happen. During sexual reproduction, sperm and egg unite to form the offspring. However, due to meiosis, the genetic combinations within those gametes are different from the parents. This is what encourages genetic diversity in a species. It is unknown, however, which of the mother's genes and which of the father's genes will be passed onto the child. It is all by chance. Mendel was able to figure out that sometimes there is a greater chance than others of certain traits being passed on. Another advantage of Mendel doing so many experiments and having so much data is that he was able to see patterns of probability emerge.

Mendel's Laws

Through his work, Mendel studied several different traits of peas. Seed color, pod location, flower color, seed and pod texture, and stem height were just a few of the traits he investigated. By pollinating the plants himself, he was able to select for one particular trait at a time and observe the outcomes. Mendel found that two "factors" governed each trait, one from each

parent. Traits or characteristics came in several forms, known as **alleles**. He established that alleles may either show up more often or show up less often.

He called the alleles that showed up more often **dominant** and those that showed up less often **recessive**. After repeated successful tests, he established his Law of Dominance, which states that in an organism with two different alleles for a trait, one will be expressed as dominant (unless the individual has two copies of the recessive allele).

Through his work, Mendel also postulated two other laws based on all of his data. He called them the Law of Segregation and the Law of Independent Assortment.

The **Law of Segregation** states that only one of the two possible alleles from each parent is passed on to each offspring. Thus the two alleles for each trait from each parent segregate into different gametes.

When parent organisms are crossed to investigate a single trait, it is called a monohybrid cross ("mono" means one). The tool geneticists use to show a cross is called a Punnett square. In a Punnett square, one parent's alleles are placed at the top of the box and the other parent's on the side. It is common practice to put the female at the top and the male on the side, but in reality it does not matter, as long as you know which is which. Another question that often arises is which letters should be used to represent the alleles. A good rule of thumb here is to choose letters where the uppercase and lowercase look different. For example, G and g, R and r, and A and a are good to use. F and f and S and s are not as clear and may cause confusion. Like the positioning of the alleles along the edges of the Punnett square, it does not matter which letters you use, as long as you are able to keep them straight.

Once the alleles have been segregated, they combine in the squares just like numbers are added in addition tables. Since a monohybrid cross only deals with one trait, and each trait is represented by two alleles, the Punnett square should have four boxes. This Punnett square shows the result of the cross of two plants that had one dominant allele and one recessive allele.

In a cross, the parent (P) generation gives rise to the F_1 (first filial, meaning "child") generation. If individuals from the F_1 generation are crossed, the resulting offspring are called the F_2 (second filial) generation.

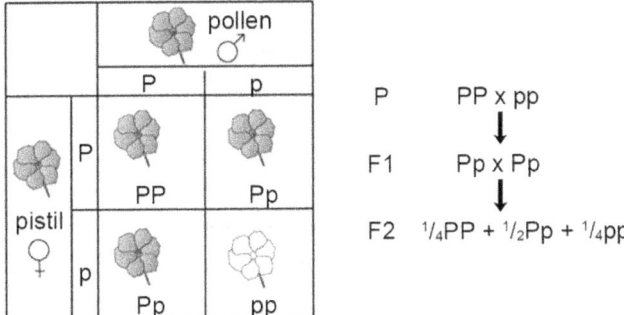

Figure 6.1. Result of cross between two plants, each with one dominant and one recessive allele

This cross results in a 1:2:1 ratio of F_2 offspring. Here, the *P* is the dominant allele and the *p* is the recessive allele. The F_1 cross produces three offspring expressing the dominant allele (one *PP* and two *Pp*) and one offspring expressing the recessive allele (*pp*). Some other important terms to know:

Homozygous means having a pair of identical alleles. For example, *PP* and *pp* are homozygous pairs.

Heterozygous means having two different alleles. For example, *Pp* is a heterozygous pair.

Phenotype refers to the organism's physical appearance.

Genotype refers to the organism's genetic makeup. For example, *PP* and *Pp* have the same phenotype (purple in color), but different genotypes.

Many times after performing a genetic cross, you will be asked to write the proposed genotypic and phenotypic ratios of the offspring. Please remember that each box of a Punnett square is a possible offspring, Many times students misread the information and assume that the upper left box is child one, the upper right box is child two, and so on. Each time the parents with the reported genotypes produce a child, there is a chance ANY of the Punnett square boxes may occur.

When a dominant trait is expressed, it can either be homozygous dominant (GG, for example) or heterozygous (Gg). If an individual shows this dominant phenotype, how do you know which condition of the genotype is being expressed? A protocol called a **testcross** can help identify the genotype in this case.

To begin, cross the unknown individual with a known homozygous recessive individual (gg). See the Punnett square below.

	G	G
g	Gg	Gg
g	Gg	Gg

In this example, the unknown genotype is homozygous dominant. When crossed with a homozygous recessive individual, all of the offspring will express the dominant phenotype no matter how many offspring are produced.

If the unknown genotype is heterozygous, however, then the recessive allele is present and would be expressed in some of the individuals. See the Punnett square below.

	G	g
g	Gg	gg
g	Gg	gg

Here, 50% of the offspring are homozygous recessive. The only way the recessive trait can be expressed is if two recessive alleles are present.

Mendel's other law is called the **Law of Independent Assortment**. It states that alleles assort independently of each other. The law of segregation applies for monohybrid crosses (only one character, for example flower color, is experimented with). In a dihybrid cross, two characters are explored. Many times traits are carried on the same chromosome, or if on separate chromosomes, they are inherited together. The inheritance of alleles is completely random, so one needs to determine what the probabilities are that certain ones will be passed on.

Remember that every trait is coded for by two alleles, one from the mother and one from the father. In a monohybrid cross, you can visualize this with a 4-box Punnett square, but since a dihybrid cross involves the inheritance

of two traits at the same time, the number of boxes in the Punnett square increases. Since each parent now has a possible four alleles (two for each trait) to pass on, the Punnett square must have 16 boxes.

When doing practice problems that involve dihybrid crosses, the key is to be certain that there is a representative allele for each trait above or next to each box. Also, after doing the cross, there should be four alleles inside of each box as well. Study the sample problem below.

Two of the seven characters Mendel studied were seed shape and color. Yellow is the dominant seed color (*Y*) and green is the recessive color (*y*). The dominant seed shape is round (*R*) and the recessive shape is wrinkled (*r*). A cross between a plant with yellow round seeds *(YYRR)* and a plant with green wrinkled seeds (*yyrr*) produces an F_1 generation with the genotype *YyRr*. (Consider this: The only alleles the yellow/round parent had to pass on were dominant, *Y* and *R*. The only alleles the green/wrinkled parent had to pass on were recessive, *y* and *r*).

Members of the F^1 generation were then crossed. Independent assortment of the *YyRr* genotype yields four possible outcomes (*YR, Yr, yR,* and *yr*). Remember, there needs to be an allele for each possible trait present. Crossing the F_1 generation would result in the production of F_2 offspring with a 9:3:3:1 phenotypic ratio.

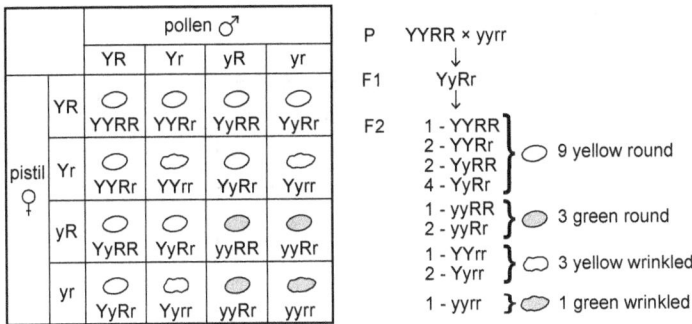

Figure 6.2. Punnet square representing an F1 cross between plants heterozygous for 2 traits

Punnett squares are a visual model used to determine the possible inheritance patterns of offspring. Also remember that each time parents reproduce, *any* of these genotypes are possible. The first offspring could be YYRr and the second offspring could be yyrr. It is completely random.

Exceptions to the Rule

Based on Mendelian genetics, the more complex hereditary pattern of **dominance** was discovered. In Mendel's law of segregation, the F_1 generation has either green or yellow seeds. This is an example of **complete dominance**. However, sometimes there are cases when one allele does not completely mask the other. Instead, a combined phenotype appears. This is called **incomplete dominance**. Here, the F_1 generation results in an appearance somewhere between the two parents.

This is quite common in a species of flower called a snapdragon. This flower produces red flowers (R) and white flowers (r). When a red is crossed with a white, the resulting offspring are pink. See the Punnett square below.

	R	R
r	Rr	Rr
r	Rr	Rr

R = red
r = white

When members of the F_1 generation are crossed with each other, the following occurs:

	R	r
R	RR	Rr
r	Rr	rr

The F_2 generation produces all possible phenotypes (red, pink, and white), in the ratio of 1:2:1.

Another exception to Mendel's law of dominance can be shown in **codominance** (also called multiple alleles). Here, the genes may form new phenotypes and are often expressed at the same time. One allele is not dominant over another, for example, in human blood types. Humans have specific antigens on their blood that mark them as either A, B, AB, or recessive O (that designates no antigens). Since all traits need to be expressed

by two alleles, a person who is homozygous for blood type A would be AA. Being heterozygous for blood type A would be AO (remember that O is the absence of an antigen). Blood type B can be either BB or BO. Antigens A and B are of equal strength and O is recessive. Type AB blood has the genotype AB. This is where the codominance comes into play. Blood type O blood has two recessive O genes.

Family History

A family **pedigree** is a collection of a family's history for a particular trait. As you work your way through the pedigree of interest, you apply Mendelian inheritance theories. In tracing a trait, the generations are mapped in a pedigree chart, similar to a family tree but with the alleles present. In a case where both parents have a particular trait and one of two children also expresses this trait, then the trait is due to a dominant allele. In contrast, if both parents do not express a trait and one of their children does, that trait is due to a recessive allele. An example of a pedigree is shown below.

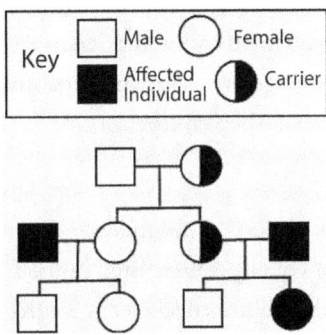

Figure 6.3. Example of a pedigree

A carrier is an individual who has the alleles for the trait, but does not express it. In order for recessive traits to be expressed, the organism needs to have a double dose of the gene. A carrier only has one. When this organism reproduces there is a chance it will pass on its recessive gene to its offspring. If the other parent also has the recessive allele, then the offspring will most likely express the trait.

More Exceptions to the Rule

Non-Mendelian inheritance is a general term describing any pattern of genetic inheritance that does not conform to Mendel's laws or does not rely

on a single chromosomal gene. These are often called polygenic inheritance patterns. Examples of non-Mendelian inheritance include complex traits, environmental influence, organelle DNA, and transmission bias.

Multiple Genes

Multiple genes determine the expression of many complex traits. For example, disorders arising from a defect in a single gene are rare compared to complex disorders like cancer, heart disease, and diabetes. The inheritance of such complex disorders does not follow Mendelian rules because they involve more than one gene. Even simple traits, like skin color, hair color, and height in humans are also caused by multiple genes.

Organelle Genetics

While chromosomal DNA carries the majority of an organism's genetic material, organelles, including mitochondria and chloroplasts, also have DNA containing genes. Organelle genes have their own patterns of inheritance that do not conform to Mendelian rules. Such patterns of inheritance are often called maternal because offspring receive most of their organelle DNA from the mother. New studies have shown that some organisms also receive genetic material from their father's mitochondria.

Transmission Bias

Transmission bias describes a situation in which the alleles of the parent organisms are not equally represented in their offspring. Transmission bias often results from the failure of alleles to segregate properly during cell division—an event called **nondisjunction**. During meiosis, homologous chromosomes do not separate properly. When this happens, one of the newly formed gametes receives a double dose of the gene and the other gamete gets none. If either of these cells then unites to become a fertilized embryo, the resulting offspring will have an abnormal number of chromosomes. The condition trisomy-21 that causes Down syndrome is one such disorder. Here, one gamete ended up with an extra 21st chromosome, giving the resulting offspring 47 instead of 46. Mendelian genetics assumes equal representation of parent alleles in the offspring generation.

Genes Can Be Linked

Genetic linkage is often considered a form of non-Mendelian inheritance because closely linked chromosomal genes tend to assort together, not

separately. Linkage, however, is not entirely non-Mendelian because classical genetics can generally explain and predict the patterns of inheritance of linked traits.

Genetic linkage is the inheritance of two or more traits together. In general, the transmission of a particular allele is independent of the alleles passed on for other traits. Independent inheritance results from the random sorting of chromosomes during meiosis. Genes found on the same chromosome, however, often remain together during meiosis. Thus, these linked genes have a greater probability of appearing together in offspring.

The phenomenon known as crossing over prevents complete linkage of genes on the same chromosome. During meiosis, paired chromosomes exchange genetic material creating new combinations of DNA. Crossing over is more likely to disrupt linkage when genes are far apart on a chromosome. Greater distance between genes increases the probability that crossing over will occur between the gene loci.

Human Genetic Disorders

The same techniques of pedigree analysis apply when tracing inherited disorders. Thousands of genetic disorders result from the inheritance of a recessive trait. These disorders range from non-lethal traits (such as albinism) to life threatening traits (such as cystic fibrosis).

Most people with recessive disorders are born to parents with normal phenotypes. The mating of heterozygous parents results in an offspring genotypic ratio of 1:2:1; thus 1 out of 4 offspring will express the recessive trait.

Examine this pedigree.

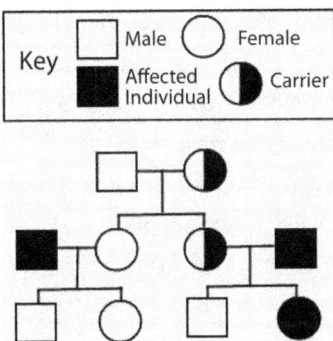

Figure 6.4. Example of a pedigree showing the inheritance of a sex-linked trait

You will remember that a pedigree shows how traits are inherited throughout a family. Many times with recessive traits, there are carriers present who do not express the trait, but can pass it along to the next generation if conditions are right. In the pedigree seen here, this family has a trait that is called sex-linked. In **sex-linked traits**, the condition is present on the X chromosome. When a male inherits this allele, he will express it. This is because males only have one X chromosome. The Y chromosome carries little genetic information relative to the X chromosome. Since men do not have a second X chromosome to mask recessive genes, recessive traits are expressed more often in men. Women must have recessive genes on both X chromosomes to phenotypically show the trait. Examples of sex-linked traits include hemophilia and color-blindness.

Sex influenced traits

Traits are influenced by sex hormones. Male pattern baldness is an example of a sex-influenced trait. Testosterone influences the expression of the gene, thus, men are more susceptible to hair loss.

Chromosome Theory

Walter Sutton introduced his chromosome theory in the early 1900s. He built on the work of the late 1800s, which defined the processes of mitosis and meiosis. Sutton understood how these processes confirmed Mendel's ideas of how "factors" were passed on. The chromosome theory basically states that genes are located on chromosomes that undergo independent assortment and segregation.

Screening for Genetic Disorders

Some genetic disorders can be prevented. Parents can be screened for genetic disorders before the child is conceived or in the early stages of pregnancy. Genetic counselors determine the risk of producing offspring that may express a genetic disorder. The counselor reviews the family's pedigree and determines the frequency of recessive alleles. While genetic counseling is helpful for future parents, it does not affirmatively ascertain whether a child will present a disorder. This type of testing is recommended for parents who have family histories of disorders or if the female is of an advanced age (over 35 years old).

There are some genetic disorders that can be discovered in a heterozygous parent. For example, sickle-cell anemia and cystic fibrosis alleles can be

discovered in carriers by genetic testing. If the parents are carriers but decide to have children anyway, fetal testing is available during the pregnancy. There are a few techniques available to determine if a developing fetus will have certain genetic disorders.

Amniocentesis is a procedure in which a needle is inserted into the uterus to extract some of the amniotic fluid surrounding the fetus. Some disorders can be detected by chemicals found in amniotic fluid. Other disorders can be detected by karyotyping cells cultured from the fluid.

A karyotype is a picture of all the chromosomes of the fetus. The genetic material is removed from the amniotic fluid and then the chromosomes are mapped out, as seen in the figures below. If there were a genetic abnormality, the karyotype would show an extra or missing chromosome (like with Trisomy-21, already described). The karyotype on the left is a normal human female. The one on the right is a female who will exhibit the symptoms of Down syndrome.

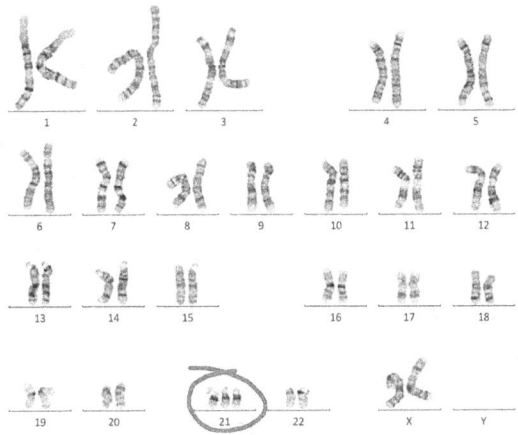

Figure 6.5. Karyotypes of a normal female fetus (left) and one that has an extra Chromosome 21 (right) and will exhibit sympotoms of Down syndrome

Karyotypes are also used to determine the sex of the child. Unlike ultrasounds, described below, a karyotype is more than 99% accurate.

A modification of the amniocentesis protocol is called **chorionic villus sampling** (CVS). Here, a physician removes some of the fetal tissue from the placenta. The cells are then karyotyped as they are in amniocentesis. The advantage of CVS is that the cells can be karyotyped immediately, unlike in amniocentesis, which takes several weeks to culture.

Unlike amniocentesis and CVS, **ultrasounds** are a non-invasive technique employed to detect genetic disorders. However, ultrasounds are limited in that they can only detect physical abnormalities of the fetus. They use sound waves to bound off of the fetus and produce an image on a computer screen. Many times doctors using this method to determine abnormalities or even the sex of the child will tell the parents that additional testing needs to be done to obtain more conclusive results.

Newborn screening is now routinely performed in the United States at birth. One disease that is screened for is phenylketonuria. Phenylketonuria (PKU) is a recessively inherited disorder that does not allow children to metabolize the amino acid phenylalanine. This amino acid and its by-product accumulate in the blood to toxic levels, resulting in mental retardation. This can be prevented by screening at birth for this defect and treating it with a special diet.

Impact of Mutations

You know that changes in an organism's genome are what will most likely cause new traits to appear. As DNA replicates, mutations occur that result in the expression of many inheritable diseases. Some of the most common genetic disorders are found in the table below.

Genetic Disorder	How it is Inherited
Phenylketonuria (PKU)	autosomal recessive
Tay-Sachs Disease	autosomal recessive
Hemophilia	sex-linked recessive
color blindness	sex-linked recessive

Those disorders that are autosomal are found in the body cells. Those that are sex-linked are found in the gametes.

Genes and the Environment

The environment can have an impact on an individual's phenotype. For example, a person living at a higher altitude will have a different amount of red and white blood cells than a person living at sea level. In some cases, a particular trait is advantageous to the organism in a particular environment. Sickle-cell disease causes a low oxygen level in the blood, which results in red blood cells having a sickle shape. About one in every ten African-Americans has the sickle-cell trait. Heterozygous carriers are usually healthy compared to homozygous individuals who suffer severe detrimental effects. In tropical African environments, heterozygote people are more resistant to malaria than people who do not carry any copies of the sickle-cell gene.

Chapter 7: Evolution

How did life begin? Why are there so many different forms of living things? While there is no one answer to these questions, there is sufficient evidence to suggest some plausible ideas. Evolution is the theme that unifies all of biology. Without it, there would be no genetics, no adaptations, no reproduction, and no biodiversity. While sometimes considered a "controversial topic," evolution is the one thing on which biologists rely for explanations of the origins of life. It is important to remember that individuals do not evolve. The changes that make species more able to survive and reproduce are evident at the population level of organization. A population is all the members of the same species living in the same area at the same time.

How do scientists know that evolution has taken place? Well, there are several different things they use to provide evidence. These can include

- DNA and molecular similarities
- body structures
- similarities in embryological development
- plate tectonics and biogeography
- the fossil record

DNA and Molecular Similarities

All living things have DNA, or some form of genetic material, that carries the blueprint for all the traits the organism expresses. This material is composed of the same code no matter in which organism you look. The nitrogen bases (remember A, T, C, and G?) in an elephant's DNA are exactly the same as the nitrogen bases found within an *E. coli* bacterium or a piece of moss. They are just arranged in a different order.

It has been found that the more closely related two species are, the more similar their genetic code is. For example, if you were to look at the genetic make up of a human and that of a chimpanzee, you would find that 99% of the code is exactly the same! The 1% difference is what makes a chimp a chimp and a human a human.

There is another molecular trait scientists examine when determining evolutionary relationships. Cytochrome c is an amino acid that is common among many organisms that carry out aerobic respiration. Like the DNA sequence, the more similar the order of bases in cytochrome c, the more closely related the species are. Human and chimpanzee cytochrome c sequences are identical. Human and giraffe sequences are very different (but still show a few similarities).

Body Structures

Scientists examine the physical properties of different organisms to determine their structure and function. Additionally, these physical properties are used to see which organisms may be related. For example, the bone structure in the wing of a bird, the human arm, the leg of a dog, and the fluke of a whale all appear to be the same (see Figure 7.1). You can see the similar bones highlighted.

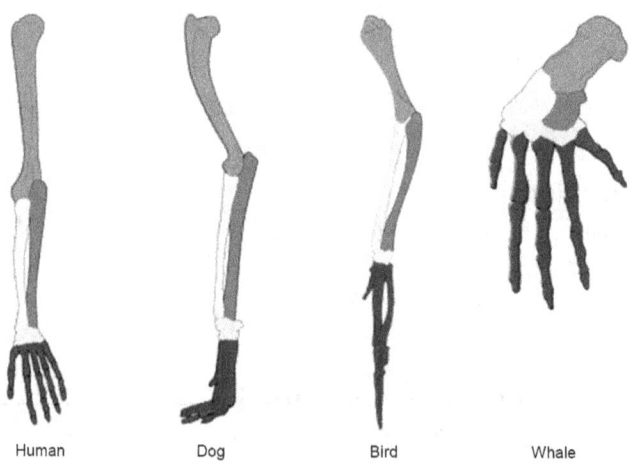

Figure 7.1. Homologies

These structures are called homologies. Why would all of these different animals have the same bones in their appendages if they did not share a common ancestor?

The opposite of a homology is called an analogy. Analogous structures are those that have the same function, but have differing structures and development. For example, the eyes of an octopus and those of a chimpanzee are both very sophisticated and well adapted for sight. Other than both organisms sharing a common animal ancestor, however, there is really very little they have in common. This suggests that eyes evolved independently in each species due to similar environmental pressures.

A final piece of structural evidence that supports evolution is called a vestigial structure. This is a structure that still exists in a modern organism but really serves little to no purpose. An example of a vestigial structure in humans would be the appendix. While people can live quite well without it, it is believed that human ancestors used this organ to help digest food.

Similarities in Embryological Development

In the late 1800s a scientist by the name of Ernst Haeckel discovered that many organisms looked very similar during the early stages of their development. In fact, he saw that the more closely related two organisms were, the more traits they shared throughout their lives. He compiled a series of drawings to show his research (Figure 7.2), each line showing further growth and development.

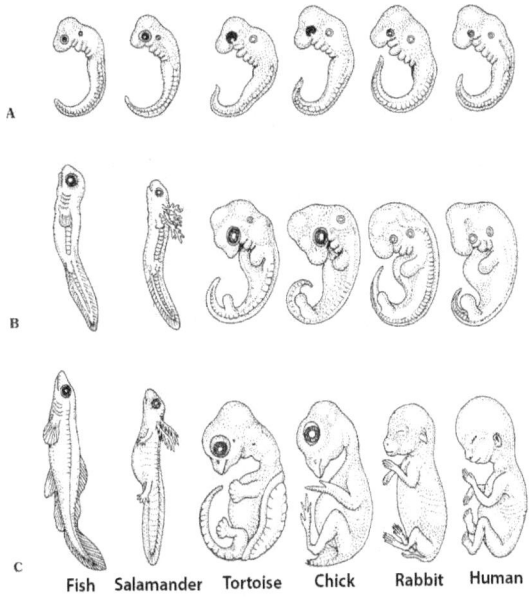

Figure 7.2. Similarities in Embryology

While these drawing have come under fire in recent years as to their accuracy, they certainly do provide support for the fact that similar organisms do share embryological traits. All of the organisms here are vertebrates. Consider all organisms in line A. If you were given diagrams of these without labels, you would be hard pressed to tell the difference. It is only when the organisms are more developed (Line C) that they start to look like their adult forms.

Plate Tectonics and Biogeography

Plate tectonics is the theory that the land masses of the Earth are found on giant plates that are slowing moving around the surface of the planet. It is strongly believed that all of the continents were once connected to each other in one giant landmass called Pangaea. As the plates moved, they carried the land and all the organisms living on that land with them. Fossil evidence has been found that shows migration routes of many species.

Fossil Record

Perhaps the most convincing evidence that supports the theory of evolution is found in the fossil record. While incomplete, there have been enough remnants of ancient organisms found that a fairly accurate picture of how life began has been created. Almost all of the organisms that have ever lived on Earth are now extinct. These dead organisms have left evidence about their existence that scientists can use to reconstruct what things were like when they were alive.

The most important fossils scientists can find are those of transitional species. These organisms show the link between modern species and those that are believed to be their ancestors. For example, it is strongly believed that modern humans evolved from apes. In 1974, fossil evidence that connected the two species was discovered in a single skeleton. Nicknamed "Lucy," this species of hominid is called *Australopithecus*. The bones suggest that she was about the size of a large chimpanzee but walked upright. Since then, many additional fossils have been found that paint quite a picture of human evolution.

The Ideas of Evolution

While there have been many ideas about how life began on Earth over the years, the two that are probably the most prominent are that of use and disuse and natural selection.

Lamarck and "Use and Disuse"

Jean-Baptist Lamarck (1744-1829) was a French naturalist who was the first person to report that traits got passed from parents to offspring. He called this the "inheritance of acquired characteristics." Lamarck came up with the idea that if an organism had a need for a certain trait then it would be deemed beneficial. This trait could then be passed along to its offspring, where it would also be beneficial. For example, a giraffe is born with a short neck. Since it needs to reach the leaves located at the tops of the trees, it has to stretch its neck to get them. Over time, the giraffe's neck would be much longer. When this long-necked giraffe reproduced, it would pass on this trait to its young. If all the giraffes were stretching their necks, then ultimately all giraffes would have long necks.

Today, some might consider Lamarck crazy, but in his time, this idea was widely accepted.

Darwin and Natural Selection

Charles Darwin (1809-1882) is probably the most famous biologist. His ideas about how species came to be have become the foundation for all of evolutionary theory. Darwin started his life pursuing a career in the church, but after a short time found that it was not his path. A former professor of his recommended him for the position of naturalist on an around-the-world voyage getting set to sail. At age 22, Darwin boarded the H.M.S. *Beagle* and spent the next five years traveling around the world. He visited many places and collected many different species.

It was when Darwin visited a small chain of islands off the coast of South America called the Galapagos that he started to put together his ideas about evolution. He had read a book by Charles Lyell called the *Principles of Geology* that proposed that the physical Earth could change. After visiting the Galapagos, he asked himself why couldn't the living Earth change as well. He noticed that each island in the chain had similar, but noticeably different species.

On returning to England, Darwin continued his research and ultimately put together all of his ideas about how organisms can change over time when the environment changes. He presented his ideas to his fellow naturalists, many of whom were awestruck by his work. While they encouraged him to publish his work, he resisted because he knew there would be a large public outcry (remember, the church was very important to people in Darwin's

day). It was not until Darwin received a letter from a young student named Alfred Russell Wallace that he decided to change his mind and go ahead and publish his findings.

In the letter, Wallace outlined his work about how species can change over time. He had used a lot of Darwin's ideas (and credited him accordingly) but was getting ready to present them to the world. Not wanting to lose credit for work he had done, Darwin got busy writing and in 1859 published *On the Origin of Species by Means of Natural Selection*. It sold out completely on the first day.

Natural selection (or "survival of the fittest" as one reporter of the day called it) is the idea that those organisms that are best fit for the environment will survive to reproduce. Those that are not fit will die. In order for natural selection to occur, the following must be in place:

1. More species must be produced than can survive.
2. There must be variation among the members of the species.
3. Since resources are in short supply, there will be competition.
4. Those organisms that have traits, called adaptations, that allow them to outcompete others will survive to reproduce. Those without those traits will die.

Now, how does variation occur? Well, the primary way is through mutation. A mutation is an alteration of an organism's DNA that produces a new trait.

Let's consider how natural selection applies to the origin of the giraffes discussed earlier. According to Darwin, there would have been many giraffes born in a population (Principle 1). Among these giraffes, there would be those with long necks, those with medium necks, and those with short necks (Principle II). The giraffes would fight each other for the leaves that were up in the trees (Principle III). Since the giraffes with long necks were better able to reach the leaves high up, they were able to outcompete those giraffes with shorter necks that would go hungry. The long-necked giraffes reproduced and made more long-necked giraffes. The giraffes with short necks did not have any food, so they became weak and died. The genes for having a short neck died with them.

Please remember, this did not happen overnight. This type of process takes hundreds, if not thousands of years.

Since Darwin first presented these ideas, there has yet to be a piece of evidence found that refutes anything he said.

Types of Natural Selection

Natural selection changes the frequency of different alleles within a population.

Stabilizing selection weeds out those genotypes and phenotypes at the extreme ends of the population. Those that are too large or too small, for example, are often dismissed in favor of more moderate sizes.

Disruptive selection is just the opposite of stabilizing selection. Here, the more extreme cases increase while the more moderate ones decrease. Sometimes two different phenotypes can exist at the same time. This can ultimately lead to the formation of a new species if conditions are right.

Directional selection occurs when one phenotype replaces another. Due to changes in environmental conditions, darker colored individuals of a species may be better at hiding than those that are lighter colored. If such a case exists, the darker colored ones would survive to reproduce. However, if the conditions changed, and the lighter ones were once again more fit, then the population would change back.

Changes in Allele Frequency

As already mentioned, the major factor that causes changes in populations to occur is a mutation. These can happen at random or be the result of some external force (such as radiation). Random mutations often occur during meiosis, when chromosomes are duplicating and their pieces are being exchanged during crossover. If a mutation occurs, then the resulting offspring will possibly express a new trait. If this trait happens to be beneficial to the current environmental conditions, it will probably survive to be passed on to subsequent generations.

There are other methods through which allele frequencies within a population can be changed. The first, called **genetic drift**, happens by chance. The occurrence of natural disasters can reduce the size of a population. This results in certain genes being more or less frequent than in the population before the event. The other way genes move into or out of a population is called **gene flow**. When individuals move into and out of a population, their presence or absence introduces or takes away specific genes that may or may not be beneficial.

How Do New Species Form?

There are several mechanisms that can cause the formation of a new species. All of them have to do with the ability of members of the population to reproduce. If something disrupts this ability, then the potential for new genes to be added to the gene pool exists.

Geographic isolation happens when there is some physical barrier between members of an existing population. This barrier may be a river, a mountain range, or even a road. Over time, the individuals on either side of the barrier become different enough from each other that they can no longer reproduce. This makes members of each population a new species.

Behavioral isolation also can lead to the formation of new species. This occurs when the behavior of one organism changes for some reason. Perhaps it starts feeding later in the day or starts to produce a different mating call. As a result, the members of the population will come into contact with each other less and less.

When **reproductive isolation** occurs, related species are unable to mate because of incompatibilities in reproductive anatomy.

Temporal isolation occurs when species are active at different times. This may be a daily cycle, or the female coming into estrus at a different time of year. If suitable mates are not found at the right time, reproduction cannot occur.

Patterns of Evolution

Convergent evolution happens when unrelated species become more similar due to similar environmental pressures. Sharks and dolphins look very much alike. They both have elongated bodies and are excellent swimmers. Dolphins are mammals while sharks are fish. They do share a common vertebrate ancestor, but that is about all they have in common.

Divergent evolution happens when similar species become more different due to environmental conditions. These conditions cause each species to eventually lose the ability to reproduce with each other. The aforementioned bat wing, human arm, and whale fluke are examples of what happens when divergent evolution occurs.

Coevolution can happen when two species are so closely interconnected that they become totally dependent upon each other. For example, hummingbirds are excellent pollinators of flowers. Their bills are especially

adapted to fit down into the throat of the flower to get the nectar they desire. The flower, in turn, attaches its pollen to the hummingbird's feathers. As the hummingbird moves from flower to flower, the pollen gets dispersed. As the flowering plant evolves, so does the hummingbird species. There are many examples of insects as well as hummingbirds that are specifically adapted to pollinate one or just a few plants (see Figure 7.3).

Figure 7.3. Coevolution between Hummingbird and Flower

Different Ideas about Evolution

While natural selection is the most commonly and widely accepted theory about how life began, there are some other ideas that provide good points for discussion.

The first is called gradualism. This idea suggests that the changes that cause the formation of a new species occur at a slow, predictable rate over time. Big changes occur as a result of many smaller ones adding up.

If this idea were true, then the fossil record should show evidence from every single organism and every trait that has evolved. As you know, however, the fossil record is quite incomplete and is also lacking many of the intermediate species that would fill in the examples that support the idea of gradualism.

Another idea of how species came to be is called punctuated equilibrium. Here, the idea is that new species appear suddenly, without prior warning, after very long periods of time without any new ones showing up.

The final idea is called spontaneous generation. This is the idea that life comes from non-life. For example, leaves falling into a stream become fish and flies emerge from rotting meat left open to the air. While a good idea at first, spontaneous generation has been disproven over and over again. The noted microbiologist Louis Pasteur was the able to show that all life needs to come from other living things. He boiled a flask containing nutrient broth and placed it on a shelf in his laboratory for one year. The flask was open to the air, but had a curved neck so any particles entering it would get trapped. After one year, the broth was still clear. When he broke the neck off the flask (allowing particles to drop right into the broth) a culture grew within a day or so.

From Life to Death

The hypothesis that life developed on Earth from nonliving materials is the most widely accepted idea. The transformation from nonliving materials to life had four stages. The first stage was the nonliving (abiotic) synthesis of small monomers such as amino acids and nucleotides. In the second stage, these monomers combine to form polymers, such as proteins and nucleic acids. The third stage was the formation of protobionts, droplets containing proteins or nucleic acids surrounded by a membrane-like structure. The last stage was the origin of heredity, with RNA as the first genetic material.

The first stage of this idea was hypothesized in the 1920s. A. I. Oparin and J. B. S. Haldane were the first to hypothesize that the primitive atmosphere was a reducing atmosphere without oxygen. The gases were rich in hydrogen, methane, water, and ammonia. In the 1950s, Stanley Miller proved Oparin's idea in the laboratory by combining the above gases. When given an electrical spark, he was able to synthesize simple amino acids. It is commonly accepted that amino acids appeared before DNA. Other laboratory experiments have also supported the other stages in origin of life theory.

Other scientists believe simpler hereditary systems originated before nucleic acids. In 1991, Julius Rebek was able to synthesize a simple organic molecule that could replicate itself. According to his idea, this simple molecule represents a precursor of RNA.

Prokaryotes are the simplest life form. Their small genome size limits the number of genes that control metabolic activities. Over time, some prokaryotic groups became multicellular organisms for this reason. Prokaryotes then evolved to form complex bacterial communities in which each species benefited from one another.

The **endosymbiotic theory** of the origin of eukaryotes states that eukaryotes arose from symbiotic groups of prokaryotic cells, that is, groups of prokaryotic cells living together. According to this theory, smaller prokaryotes lived within larger prokaryotic cells, eventually evolving into chloroplasts and mitochondria. Chloroplasts are the descendants of photosynthetic prokaryotes, and mitochondria are likely the descendants of bacteria that were aerobic heterotrophs. Serial endosymbiosis is a sequence of endosymbiotic events and may have played a role in the progression of life forms that became eukaryotes.

Extinction occurs when a species ceases to existence and results in reduced biodiversity. The point of extinction is generally considered as the death of the last member of a given taxon, but determining the exact moment of extinction is not easy. In the context of evolution, new species are created by speciation—where new varieties of organisms arise and thrive when they are able to find and exploit an ecological niche. Species become extinct when they are no longer able to survive in changing conditions or against superior competitors.

Prior to the spread of human beings across the earth, extinction was a purely natural phenomenon that occurred at a low rate. A typical species becomes extinct within 10 million years of its first appearance, although some species, called "living fossils" survive virtually unchanged for hundreds of millions of years. Mass extinction was very rare. Within the last 100,000 years as the numbers of human beings have continued to grow and expand, however, species extinction has increased at an unprecedented rate.

A species becomes extinct when the last existing member of that species dies. Extinction therefore becomes a certainty when there are no surviving individuals that are able to reproduce and create a new generation. A species may become functionally extinct when only a handful of individuals survive, which are unable to reproduce due to poor health, age, sparse distribution, or other reasons.

It is important to note that humans have also made significant attempts to preserve critically endangered species through the creation of the conservation status and captive breeding programs.

Chapter 8: Diversity of Life

Scientists estimate that there are more than 10 million different species of living things. These range in size from bacteria, which can only be seen with a powerful microscope, to the blue whale, which can reach lengths 35 meters long and weigh over 99,000 kilograms. And let's not forget about the plants. The largest trees are the giant sequoias found in California. These can reach heights of 84 meters and be almost eight meters in diameter.

Most species are still unknown to man. There are many places still left to look, including the canopies of the tropical rainforests, the deep oceans, and of course, the microscopic world. Trying to keep track of the approximately 1.5 million species that have been named and classified is a challenge. Sorting through all the living things requires a methodical and detailed system. This is called **taxonomy**.

The Science of Classifying Living Things

Imagine you are in the library and want to find a book about the evolution of land plants. You go to the computer and search "evolution of land plants." On the screen you see several books that might fulfill your needs. Each book is given a number (somewhere in the 500s). You write down the specifics (or print them off) and then go to the shelf to find the books.

What kinds of books can be found on either side of the one you want? Books about baking a cake? Books about poetry? No. You will find other books about plants. This is done for a reason. It makes things easier to find. Just picture how hard it would be to find things if all the books in the library were shelved according to size!

Just as in the library, living things are classified by their characteristics. The first person to develop a system of classification was the ancient Greek

philosopher Aristotle. He classified living things based on what they looked like and where they lived. For example, there were land plants, water plants, and air plants (like vines that wrapped around trees). He had land animals, water animals, and air animals. While not very specific, it worked for the time.

Over the centuries, naturalists built on Aristotle's work but found that they needed a better way to refer to specific organisms. When describing a living thing, most people use its common name. They refer to the organism as an oak tree or as a blue whale. As language has evolved, however, scientists noticed that the same common names were being used for different organisms in different parts of the world. Therefore, an "oak tree" in one location might refer to a completely different organism in another.

Binomial Nomenclature

In the 18th century, botanist Carl von Linné designed a method of classifying all living things using a two-word naming system. This system, called binomial nomenclature, used common characteristics to group like organisms together. Most of these names are derived from the Latin and form an organism's scientific name. The first word of each name is called its genus and the second word is called the species. For example, the scientific name for the humpback whale is *Megaptera novaeangliae*. "Mega" means "big," "ptera" means "winged," "nova" means "new," and "angliae" means "England." The translation of the scientific name is the "big-winged New Englander." If you have seen a humpback whale, you know that its front fins look like wings and that they summer in New England waters. Also note that all scientific names are either underlined *or* italicized and the first letter of the genus is capitalized. Since most scientific names are based on Latin, Linné gave himself the scientific name Carolus Linnaeus. He is considered the father of taxonomy.

Does King Phillip Come Over For Good Science?

Linnaeus took the naming a bit further and created the system that is still in use today. Realizing that many organisms were related to each other, albeit distantly, he devised a way to classify every living thing. It is important to remember that taxonomy is fluid. As new discoveries are made (especially with DNA technology), names and classifications change. Organisms that were once thought to be very closely related turn out to be not related at all and vice

versa. Therefore, take the following information with a grain of salt because it may change soon, or you may have learned it differently in your biology class.

The largest grouping of organisms is called the Domain. There are currently three domains into which every living thing falls:

- **Domain Archaea:** These are the prokaryotes that live in extreme environments. They can be found living inside of volcanoes, inside the stomachs of cows, and in areas of extreme saltiness.
- **Domain Bacteria:** These are the prokaryotes that live everywhere else. They are microscopic and can be found on every surface, inside of your intestines, and even in the Antarctic waters.
- **Domain Eukarya** This classification encompasses everything else. All organisms here have eukaryotic cells. This domain can be further divided into the protists, fungi, plants, and animals.

The rest of the classification scheme Linnaeus devised goes like this:

Eukarya — Kingdom — Phylum — Class — Order — Family — Genus — species

For purposes here, there are four kingdoms. There have been other models that include five, six, and even eight kingdoms (see Figure 8.1).

The four kingdoms are:
- Protista
- Fungi
- Plantae
- Animalia

As you move through the other levels of classification, the number of different possibilities increases.

Let's use the humpback whale example to see how this works.

Domain – Eukarya

Kingdom – Animalia

Phylum – Chordata

Class – Mammalia

Order – *Cetacea*

Family – *Balaenopteridae*

Genus – *Megaptera*

Species – *novaeangliae*

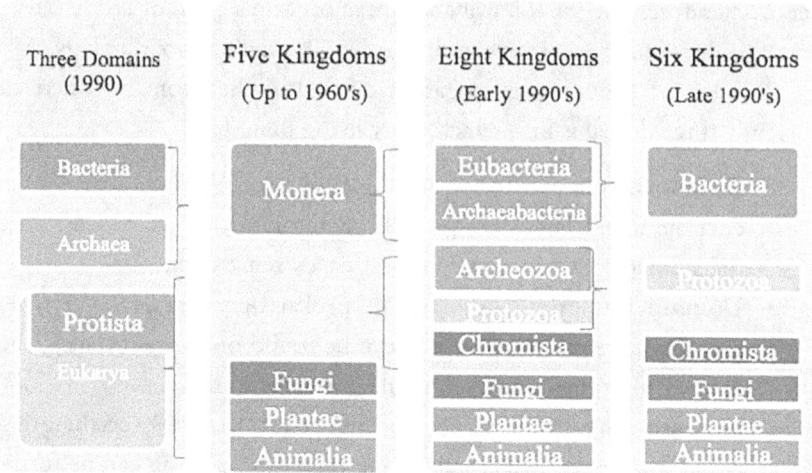

Figure 8.1. Changes in Taxonomic Kingdoms

So you have an idea of which animal is being discussed here, see Figure 8.2 for an illustration. Remember that all living things have this type of classification scheme.

Figure 8.2. Humpback Whale (*Megaptera novaeangliae*)

There are many other animals. There are many other chordates (animals with backbones). There are many other mammals (dogs, lions, seals). Whales and dolphins make up the cetacean order. Baleen whales (no teeth) can be found in the Baleanopteridae. But there is only one *Megaptera novaeangliae*. Think back to the title of this section. You may be wondering what King Phillip has to do with taxonomy. Well, the best way to learn the levels of classification in order is to create a pneumonic device. Relate each level to the first letter in the phrase "Does King Phillip Come Over For Good Science" and you will never forget which levels are most inclusive and which are most specific. "King" = Kingdom, "Phillip" = phylum, and so on.

The Diversity of Life – An Overview

Trying to do a complete analysis of all the different life forms in this review book would increase its page length to well over 5,000. Since you do not have the time, nor want, to read all of that, this survey will highlight the important parts of each of the major classifications of organisms.

Viruses

Viruses are not considered living things, so they are not classified into any of the three domains. Yes, they adapt (which is why you need to get a flu shot every year) and they can reproduce, but they do not exhibit the other characteristics of life. Since they greatly affect other living things, they need to be briefly discussed here. Viruses work by disrupting cell activity. They are obligate parasites because they rely on the host for their own reproduction. Viruses are composed of a protein coat and a nucleic acid, either DNA or RNA. A **bacteriophage** is a virus that infects a bacterium. It basically consists of a strand of genetic material surrounded by a protein coat (Figure 8.3). Animal viruses are classified by the type of nucleic acid, presence of RNA replicase, and presence of a protein coat.

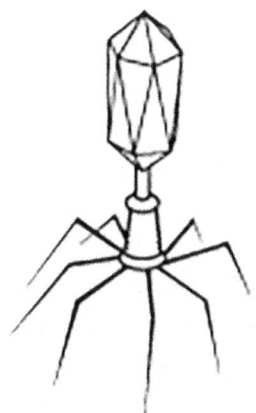

Figure 8.3. Bacteriophage

There are two types of viral reproductive cycles:

- **Lytic cycle:** The virus enters the host cell and makes copies of its nucleic acid and protein coat, and then reassembles. Afterward, it lyses or breaks out of the host cell and infects other nearby cells, repeating the process.

- **Lysogenic cycle:** The virus may remain dormant within the cell until some factor activates and stimulates it to break out of the cell. Herpes is an example of a lysogenic virus.

Domain Bacteria

Let's start with the most abundant grouping of organisms there is. Bacteria can be found everywhere. Bacteria lack a defined nuclear membrane and other bound organelles (such as mitochondria, chloroplasts, and Golgi bodies) but do contain circular chromosomes and ribosomes. The ribosomes of bacteria are much smaller than those of eukaryotes. The chromosome of Bacteria is usually a single, circular molecule that does not combine with histones.

Other characteristics of the Domain Bacteria include:

- Most bacteria possess a cell wall made of **peptidoglycan**, a combination of sugars and proteins. This cell wall helps scientists to identify different bacteria depending on which color it stains.
- All bacteria are single celled. Sometimes they clump together into colonies, but each cell retains it own functionality.
- All bacteria are **prokaryotic.** This means they do not have a membrane-bound nucleus or other organelles.
- Bacteria reproduce using either **binary fission** or **conjugation**. Binary fission involves copying the genetic material and then splitting in half. Conjugation involves the transfer of genetic material from one bacterium to another through a conjugation tube. This process increases the genetic diversity of the individual.
- Bacteria can be either autotrophic or heterotrophic.
- Some bacteria—called **pathogens**—cause diseases.
- Some bacteria are helpful. Many are **decomposers**, breaking down and recycling nutrients from dead organisms. Other bacteria are used to make medicines and foods. The tangy taste in yogurt comes from the live bacterial cultures in it.
- Some bacteria use cilia or flagella for locomotion. **Cilia** are hair-like projections coming off of the cell membrane. A **flagellum** is a single protrusion that looks like a tail coming from the cell membrane.

A typical bacterial cell can be seen in Figure 8.4.

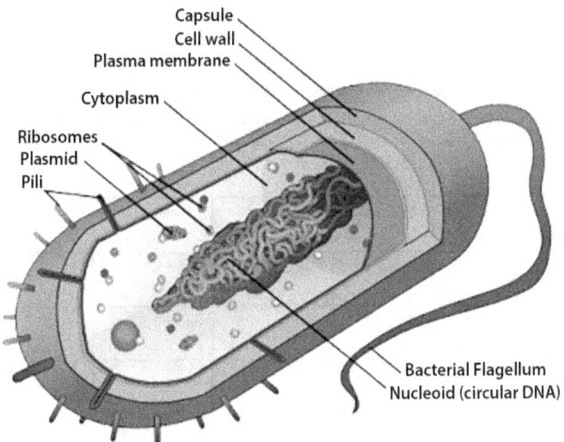

Figure 8.4. Common Bacteria

Domain Archaea

Similar to bacteria in cell structure and function except for their cell membranes and cell walls, the Archaea more closely resemble eukaryotes in the way they transcribe and translate RNA. The Archaea are single-celled and widespread, living in many environments. Many are known as **extremophiles**, thriving in places inhospitable to other life.

Examples of extremophile Archaea include

- **Extreme halophiles**, which can live in saturated salt solutions like the cyanobacteria. They live in places such as the Dead Sea and the Great Salt Lake in Utah.
- **Extreme thermophiles**, which can live only in acidic hot springs, like those in Yellowstone Park.
- **Methanogens**, obligate anaerobes that produce methane through the conversion of hydrogen.

Domain Eukarya

The Eukarya domain includes all members of the protist, fungi, plant, and animal kingdoms. Basically, if an organism is not an archaea or a bacterium, it falls into this category. Eukaryotic cells possess a membrane-bound nucleus and other membranous organelles (e.g., mitochondria, Golgi bodies, ribosomes). Also, unlike Bacteria, the chromosomes of Eukarya are linear and usually complex with **histones** (protein spools).

Table 8.1 outlines the cellular characteristics of members of the Eukarya compares them to the Archaea and Bacteria domains.

Important characteristics determining the organization of a Six-kingdom classification						
	Bacteria	Archaea	Protista	Fungi	Plantae	Animalia
Cells	Unicellular	Unicellular	Unicellular	Unicellular/ Multicellular	Multicellular	Multicellular
Nucleus	Prokaryotes	Prokaryotes	Eukaryotes	Eukaryotes	Eukaryotes	Eukaryotes
Chromosomes	Circular	Circular	Linear	Linear	Linear	Linear
Ribosome size	70 S	70 S	80 S[1]	80 S[1]	80 S[1]	80 S[1]
Feeding mode	Autotroph / Heterotroph	Autotroph / Heterotroph	Autotroph / Heterotroph	Heterotroph	Autotroph	Heterotroph
Membrane lipids	glycerol-ester lipids	glycerol-ether lipids	glycerol-ester lipids	glycerol-ester lipids	glycerol-ester lipids	glycerol-ester lipids
Cell division	Binary fission	Binary fission	Mitosis and meiosis	Mitosis and meiosis	Mitosis and meiosis	Mitosis and meiosis
Energy organelle	Absent	Absent	Mitochondria/ plastids	Mitochondria	Mitochondria/ plastids	Mitochondria
Cell wall components	Peptidoglycan (sugar + proteins)	Proteins and polysaccharides	Various	Chitin	Cellulose	Cell walls absent
Cilia/flagella	Present	Absent	Present	Present	Present	Present
Vacuoles	Absent	Absent	Absent	Present	Present	Absent

[1]Except inside the mitochondria and chloroplasts, where ribosomes are 70 S.

Kingdom Protista

Protists are eukaryotic, usually single-celled organisms (though some protists are multicellular). The Kingdom Protista is diverse, including members with characteristics of plants, animals, and fungi. All protists possess nuclei and some types of protists possess multiple nuclei. Most protists contain many mitochondria for energy production, and photosynthetic protists contain specialized structures called plastids where photosynthesis occurs. Motile protists possess external cilia or flagella. Finally, many protists have cell walls that do not contain cellulose.

The animal-like protists are called **protozoans**. Examples of these include paramecium, amoeba, and euglena. The euglena form sort of a missing link between the protozoans and the algae because they exhibit characteristics of both groups. When light is available, euglena photosynthesize. They have many

chloroplasts to collect the light and convert it into energy. However, when light is absent, euglena can become heterotrophic and actively hunt for prey.

The plant-like protists are called **algae** (commonly called seaweed). Algae are usually classified by what color pigments they contain. Common colors are red, green, and brown. A common brown alga is called kelp. It grows in very dense underwater forests and is a prime ingredient in sushi.

Kingdom Fungi

The members of this kingdom are not a lot of fun to be with at parties (Get it? Fun – guy!). Instead, they are mostly decomposers, recycling nutrients from dead organisms back into the ecosystem. Back in Aristotle's day, fungi would have been considered plants. After all, they look like plants, do not move and often come in very bright colors. It is now known that fungi have very little relation to plants and are actually more closely related to animals.

- Fungi are eukaryotic organisms that are mostly multicellular (single-celled yeasts are the exception).
- They possess cell walls composed of a polysaccharide called **chitin**.
- Fungal organelles are similar to animal organelles.
- Fungi are non-photosynthetic and possess neither chloroplasts nor plastids.
- Many fungal cells, like animal cells, possess centrioles.
- Fungi are also non-motile.
- They release exoenzymes into the environment to dissolve food from the outside. This is why they are more closely related to animals: they have to consume their food to make energy.
- The body of a fungus—called a **mycelium**—is composed of a large mass of singular strands called **hyphae** that absorb nutrients. These hyphae can stretch out for hundred of meters from the fungus.
- The reproductive structures of fungi, called **spores**, are often carried by the wind.
- Fungi can also reproduce asexually through budding.
- Common fungi include mushrooms, the mold that appears on old bread, yeast, which is used to make bread rise, and the "blue" in blue cheese.

Fungi can live in almost every habitat (the poles being the exception). They tend to like moist, dark places, which is why bread mold is so successful. The back of the refrigerator is a great place to grow. Sometimes

fungi cannot survive on their own, so they form mutualistic relationships with other organisms. **Lichens** are one such example. Lichens can often be found growing on tree trunks or on rocks. The algae photosynthesize to provide energy for the fungi, and the fungi give the algae a safe place to live. Lichens are often the first organisms to inhabit an area that has been disturbed by some catastrophic event. Volcanic eruptions and glacial retreats are such examples. The lichens are called **pioneer species** because they are the first ones to colonize an area devoid of life. Lichens attach to rocks, using their symbiotic algae for food. Over time, the lichen secretes acids that break down the rocks, which, along with the nutrients provided when the lichens die, creates soil. Small plants can then colonize the area. This continues for a long time until a forest grows. This process is called **primary succession**.

Kingdom Plantae

The review here is putting plants before animals. However, this in no way is an endorsement for the idea that plants are any less complex or important than animals. In fact, it could be argued that plants are far more complicated than animals. Sure, they do not think or have other cognitive processes, but do you remember the section that discussed photosynthesis and cellular respiration? You, as an animal, only perform cellular respiration. Plants do both. If you ever saw all of the biochemistry of the Calvin cycle or the electron transport chain, you would probably agree that plants are very complex indeed.

- Plants are eukaryotic, multicellular, and have square-shaped cells with rigid cell walls composed mostly of cellulose.
- Plants do not move from place to place.
- Plant cells contain chloroplasts and plastids for photosynthesis.
- Plant cells generally do not possess centrioles.
- Plant cells have a large, central vacuole that occupies 50%–90% of the cell interior and limits the cytoplasm to a small part of the cell. The vacuole stores acids, sugars, and wastes.
- Plants have two body forms: the haploid gametophyte and the diploid sporophyte. Each one produces structures that are important for reproduction.
- Plants are often classified by how they reproduce—by cones, spores, or flowers.

Plant evolution and diversity

About 500 million years ago, primitive plants moved from a water environment to the land. Advantages of life on land included more available light and a higher concentration of carbon dioxide. Originally, there were no predators and less competition for space on land, which made it a great place for them to live. Due to these new challenges, plants had to evolve methods of support, reproduction, respiration, and conservation of water. Specific plant tissues evolved in order to obtain water and minerals from the soil and transport them throughout the plant. Many plants developed a wax cuticle to prevent the los of water while the leaves capture light and carbon dioxide for photosynthesis. **Stomata** provide openings on the underside of leaves for oxygen to move in or out of the plant and for carbon dioxide to move in. Roots evolved to provide a method of anchorage, and the polymer lignin evolved to provide structural support (see the more thorough description of these plant traits in Chapter 9).

A simplified cladogram of plant evolution is shown in Figure 8.5. Notice that there are three derived traits – vascular tissues, seeds, and flowers.

The development of vascular tissues happened very early during plant evolution. This adaptation is used as one way to classify plants.

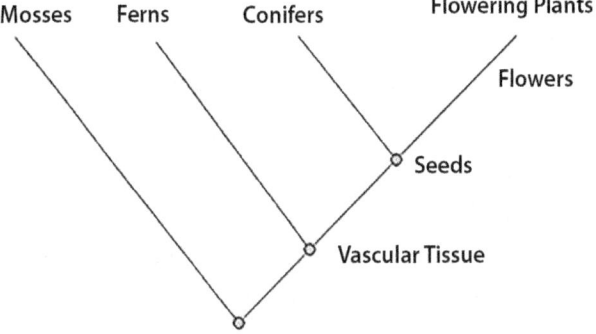

Figure 8.5. Plant Evolution Cladogram

Bryophytes are **non-vascular plants** and are characterized by several primitive features. They lack roots and conducting tissues, which increases their reliance on absorption of water that falls on the plant or condenses on the plant in high humidity. As such, they are limited to living in areas that maintain high levels of moisture at all times. They also lack leaves. Photosynthesis occurs as part of the main body of these plants. Non-vascular plants include the liverworts, hornworts, and mosses. Each is recognized as

a separate **division** (for some reason, taxonomists use the term "division" instead of "phylum" when discussing the classification of plants).

Think about where you find moss. The old adage is that moss can be found on the north side of a tree. This has helped many a lost hiker in the woods. Why does moss grow on the north side of the tree? Well, it was just established above that mosses need to live in areas that are always damp. Since the north side of a tree never gets direct sunlight, it is always going to be cooler and more damp than the other sides of the tree.

Most of the plants with which you are familiar are vascular plants. **Vascular plants** are so named because they have tubes that transport materials. **Tracheids** (also called xylem) transport water from the roots up to the leaves. **Sieve cells** (also called phloem) carry nutrients from the leaves back down to the roots for storage. Vascular plants also contain lignin, which provides rigidity and strength to cell walls for upright growth—a critical adaptation to gather sunlight for photosynthesis. Vascular plants can also have an underground stems (rhizomes) from which adventitious roots originate.

There are two kinds of vascular plants: non-seeded (club mosses, horsetails, and ferns) and seeded (the gymnosperms and the angiosperms). Remember from Figure 8.5 that the evolution of seeds is a derived (not original) character in plants. The seeded vascular plants differ from the non-seeded plants by their method of reproduction, which we will discuss later.

The club mosses, horsetails, and ferns make up the seedless vascular plants. They

- reproduce using spores.
- prefer damp, dark environments.
- use alternation of generations. In fact, the gametophyte stage of a fern's life cycle is almost undetectable due to its small size.

The largest difference between the two main groups of seeded vascular plants is the method by which they protect their seeds. A **seed** is a plant embryo that is surrounded by a protective coat. Inside this coat is a food supply that will nourish the embryo as it develops. The two main classifications of seed plants include the gymnosperms and the angiosperms.

The **gymnosperms** ("naked seed") were the first plants to evolve the use of seeds for reproduction, which made them less dependent on water to assist in reproduction. They protect their seeds within a structure called a **cone** (gymnosperms are also called the "cone-bearers" or **conifers**). Most

gymnosperms have both male and female cones. While the structure of these cones varies, they all serve to protect the sex cells and the seeds. Because the male and female gametes are located deep within their cones, pollination has to be carried out using the wind. Pollen is blown from plant to plant with hope that it will find a receptive female gamete.

Common examples of gymnosperms include the cycads, ginkgos, gnetophytes, and the true conifers. Due to the presence of small leaves on these plants, they are well adapted for survival in areas where water is limited and temperatures are low. For example, the state tree of Alaska is the Sitka spruce. It has very thin needles (modified leaves) that reduce the amount of water lost (much of interior Alaska does not get much rain).

The other group of seeded vascular plants is the angiosperms. **Angiosperms** are the largest group in the plant kingdom. They are the flowering plants and reproduce using seeds protected by an ovule. Leaf size and shape varies greatly between species of plants, and botanists often identify plants by their characteristic leaf patterns.

Angiosperms arose about 70 million years ago when dinosaurs were disappearing. The land was drying up and the plants' ability to produce seeds that could remain dormant until conditions became acceptable allowed for their success. Angiosperms also developed more advanced vascular tissue and larger leaves for increased photosynthesis. The other important adaptation present in angiosperms is the production of fruit. A fruit is a ripened ovary used to protect the seeds. It also helps with seed dispersal (see chapter 9 for a discussion of fruits and flowers).

Angiosperms can be divided into two main groups: the monocots and the dicots. A cotyledon is the leaf that is first to emerge from the seed when it germinates. **Monocots** have one cotyledon (seed leaf) that comes out, while eudicots have two. Table 8.2 outlines the major differences between these two groups of flowing plants.

Table 8.2. Comparison of Monocots and Dicots

Trait	Eudicot	Monocot
# of cotyledons	2	1
Flower parts	Groups of 4 or 5	Groups of 3
Vascular bundle arrangement	Ring	Scattered
Veins in leaf	Parallel	Fibrous

Angiosperms show the most diversity of all the plant groups. Flowers come in all different colors, shapes and sizes. Examples of monocot plants include corn, rice, and wheat grass. Examples of dicot plants include maple trees, cedar trees, irises, and tulips (a quick note here. Notice that the dicots comprise both woody and non-woody plants. All plants that produce wood are dicots, but not all dicots are woody. Remember that!).

Kingdom Animalia

Ok. You made it. Here you are at the kingdom with which you probably most familiar and even like the best—the Animal kingdom. Animals are probably the most popular with students because they are easily seen, have a large diversity, and are most like you. They behave and function in a very similar way to you, so you can relate. Most of the characteristics described here apply to humans as well.

Animals are eukaryotic, multicellular, and **motile** (this means they move around). Animals

- are all multicellular.
- are all heterotrophic.
- reproduce sexually (although there are a few exceptions).
- are classified by external, internal, and genetic characteristics, along with how they develop.

Animal evolution and diversity

Animal body plans have become increasingly complex though evolution. The development of specialized cells and tissues has allowed for complex behaviors. There is a wide range of specializations throughout the kingdom. For instance, Planarians (flatworms) have three layers of tissues that arrange themselves close together but do not form a body cavity. Annelid worms (the truly segmented worms like the earthworm) have the same three layers of tissues but do form a body cavity (called a **coelom**). Having a fluid-filled coelom provides a rigid structure against which muscles can pull (to aid in faster locomotion). It also enables more efficient collection and excretion of metabolic wastes, and is critical in developing more complex organs and larger body size.

There are three main layers of tissue that give rise to an animal's body structures. The first is called the **endoderm**, which is the innermost layer and gives rise to the to an animal's guts (mostly the digestive system). The next layer is called the **mesoderm**, which becomes the various support structures for the

animal—the muscles, skeleton, and the blood. The outmost layer is called the **ectoderm**, which gives rise to the nervous system and the external skin. Most animals arise from these three tissue layers. They are called **triploblastic**. The sponges (Porifera) are the simplest of the animals and only have one layer of tissue. They are called **monoblastic**. Jellyfish (Cnidaria) and their relatives have two layers of tissue, so they are called **diploblastic**.

Triploblastic animals can be further divided by the type of body cavity they may or may not possess. Remember that a coelom is a body cavity—basically an opening within an organism that contains its organs. The three types of body cavities are acoelomates, pseudocoelomates, and coelomates.

- **Acoelomates** have no defined body cavity. An example is the flatworm (Platyhelminthes), which must absorb food from a host's digestive system.
- **Pseudocoelomates** have a body cavity that is not lined by tissue from the mesoderm. An example is the roundworm (Nematoda).
- **Coelomates** have a true fluid-filled body cavity called a coelom derived from the mesoderm and include all "higher" animals, from annelids to insects to vertebrates. These are the animals that have developed the most sophisticated organ systems, locomotion, senses, and behavior.

Further evolutionary developments among animals trended toward more efficient locomotion, digestion, and respiration and faster responses to stimuli as more animals sought to catch an increasing variety of faster prey or to escape being eaten. The other major adaptation that led to the success of many animals was the development of a head. Called **cephalization**, the head concentrated many of the stimuli-receiving cells into the anterior end of the animal. This enabled them to focus their attention in one direction and to better escape predators.

Survey of the Animal Kingdom

As of right now, the kingdom Animalia is divided up into over 30 different phyla. Remember that taxonomy is a fluid science, and there will always be changes as new information, or new species, get discovered. For your purposes, this review will focus on the major, more common phyla. Please note that these synopses provide a brief overview of the major characteristics of each group and are, by no means, exhaustive.

Porifera: Sponges contain **spicules** for support. They possess flagella for movement in the larval stage but later become **sessile** and attach to a firm object. Sponges may reproduce sexually (either by cross- or self-fertilization) or asexually (by budding). Porifera are filter feeders and digest food by phagocytosis. They require water to support their hydroskeleton and are therefore mostly aquatic.

Cnidaria (Coelenterata): Cnidarians include marine and freshwater anemones, hydroids, corals, and jellyfish. They possess stinging cells called **nematocysts**. They may be found in a sessile **polyp** form with the tentacles at the top or in a moving **medusa** form with the tentacles floating below. Jellyfish have a hydrostatic skeleton that requires water for support. They possess no true muscles but have muscle tissue, which allows a jellyfish to jet out water from its bell to propel it upward. Cnidarians may reproduce asexually (by budding) or sexually. They are the simplest animals to possess a primitive nervous net with multifunctional neurons.

Platyhelminthes: Aptly called flatworms, these largely parasitic worms use their flat shape to aid in the diffusion of gases. They are the first group with true muscles and a bilateral nerve cord with clusters of nerve cells called **ganglia**. Flatworms can reproduce asexually (by regeneration) or sexually. Platyhelminthes may be hermaphroditic, possessing both sex organs but cannot fertilize themselves. The tapeworm is an example of a platyhelminth.

Nematoda: The roundworms are the first animals with a digestive system that contains a separate mouth and anus. Roundworms may be parasites or simple consumers. Nematoda reproduce sexually with male and female worms. They possess longitudinal muscles and thrash about when they move.

Mollusca: The clams, snails, slugs, and octopuses are soft-bodied animals, most of which are able to make a shell for protection from predators. The shell is produced by a structure called a **mantle**. The octopuses, squid, and nautilus, do not make a shell. These animals (the cephalopods) are an anomaly in the invertebrate world, having advanced vision, supersensory tentacles, and astonishingly quick and sophisticated behaviors that make them both excellent predators and astonishing escape artists. There are also many studies that support the idea of their intelligence. Many sea slugs sport brilliant warning coloring indicating that they are poisonous so they don't need protective shells. Most molluscs have a muscular foot for movement

and breathe through gills. They have an open circulatory system with sinuses bathing the body regions.

Annelida: The segmented worms (earthworm, sandworm, leech) are among the first animals with specialized tissue. The advanced circulatory system of these worms has blood vessels that operate in a closed system. They use **nephridia** as their excretory organs. Segmented worms support themselves with a hydrostatic skeleton, have circular and longitudinal muscles for movement, and ganglia at the head end specialized for light and scent detection. Annelida are hermaphroditic and each worm fertilizes the other upon mating.

Echinodermata: The sea urchins and starfish are marine animals that have spiny skins. They exhibit radial symmetry (body structured around a hub like a bicycle wheel). They possess an internal system of tubes that takes in water from the ocean and pumps it around their bodies. They use this water both to aid in support, and to power their tube feet. As echinoderms channel water into different parts of their bodies, the suction-equipped tube feet extend or retract, enabling movement. Echinoderms are excellent predators, feeding on clams, mussels, and other bivalves.

Arthropoda: This enormous phylum—the largest in the animal kingdom—includes insects, crustaceans, and spiders. Members of this group all have jointed legs. Phylum Arthropoda accounts for about 85% of all the animal species. Arthropoda possess an exoskeleton made of **chitin**. Arthropods must molt to grow, which means they shed their shells and grow new ones. They breathe through gills, trachea, or book lungs. Movement varies with members being able to swim, fly, and crawl. There is a division of labor among the appendages (legs, antennae, etc.). This is an extremely successful phylum with members occupying diverse habitats.

Chordata: All animals with a notochord or a backbone are chordates. The classes in this phylum include Agnatha (jawless fish), Chondrichthyes (cartilage fish), Osteichthyes (bony fish), Amphibia (frogs and toads; possess gills that are replaced by lungs during development), Reptilia (snakes, lizards; the first to lay eggs with a protective covering), Aves (birds; warm-blooded), and Mammalia (warm-blooded animals most of which do not lay eggs and possess mammary glands to produce milk for their young).

Mammals are characterized by how they reproduce. Most mammals—called **placental**—give birth to young that can survive outside the womb. However, as you know, in biology there are exceptions to everything. Another group of

mammals—the **marsupials**—are the animals (like kangaroos and opossums) that have a pouch. The baby is born without being fully developed. It then crawls through its mother's fur until it reaches her pouch where it latches onto a nipple inside of the pouch and stays there until it is fully formed. The third type of mammal—the **monotremes**—is an egg-layer. Yes, you read that correctly—egg layers. Their young hatch from eggs and then suckle as they grow. There are only two species in this group, the echidna and the platypus. Both live in Australia.

Evolutionary History

To make sense of the variety of anatomy and physiology in nature, it helps to consider a tool called a **phylogenetic tree** (Figure 8.6). This diagram is designed to point to common ancestry and traits between organisms. For instance, Figure 8.6 shows that all vertebrates are related to each other. Characteristics they all share include a backbone, and similarities in embryology. The typical graphic representation of classification in a phylogenetic tree depicts the hypothetical relationships between organisms within a group based on branching of lineages through time. Every time you see a phylogenetic tree, you should be aware that it is making statements about the degree of similarity between organisms, or the particular pattern in which the various lineages diverged (phylogenetic history).

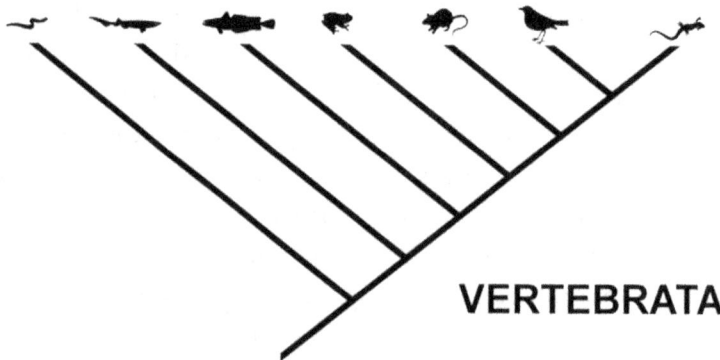

Figure 8.6. Phylogenetic Tree

Cladistics is the study of the phylogenetic relationships between organisms through analysis of shared, derived characters. The presence of a characteristic in two groups that was also present in a common ancestor (but not in older ancestral species) is considered evidence that the two groups are closely related. Cladograms are constructed to show evolutionary pathways.

Chapter 9: Organismal Biology

Multi-cellular organisms are a showcase of evolution, expressing a mindboggling array of ways to survive, cope with adversity, and reproduce. Plants and animals demonstrate countless options in body plans and structures; all function, however, to provide the organism with energy and enable it to grow and pass on its genes. This chapter discusses the anatomy, physiology, reproduction, and development of plants and animals.

Plant structure and function

Let's start the review of structure and function with plants. Yes, you may be saying that plants are boring. Based upon what you have studied so far, hopefully you realize that plants are just as, if not more, complicated organisms than animals. Sure they do not have brains, but remember from the cells and energy chapter that plants perform both cellular respiration *and* photosynthesis. You, as an animal, only perform cellular respiration. This makes plants far more complicated in my book.

Ever since plants moved onto land, they have had to overcome the obstacles of taking in sunlight and absorbing water. Since they are unable to move about, they had to evolve a variety of different structures and behaviors to allow them to not only make energy for themselves, but also to reproduce successfully. Roots, stems, and leaves are the most functionally important parts of plant anatomy.

Roots

Some people might argue that plants are built upside down. After all, the rain falling from the sky needs to get into the leaves. It seems counterproductive for it to have to travel all the way up from the ground to get there. **Roots** absorb water and minerals and exchange gases in the soil.

Roots cover a lot of area. If you have ever tried to dig up a full-grown tree, you know that the root system can spread out for great distances. Also, roots are covered with tiny projections called root hairs. These hairs increase the surface area of each root, allowing it to take in more water and minerals. In addition to water and mineral absorption, roots also anchor plants in place. This keeps the plant from moving and prevents erosion of the soil.

Roots come in one of two different forms. A **taproot** is a single, thick root that extends from the bottom of the plant. It then branches out into smaller lateral roots. Many plants use this kind of root system for nutrient storage. Carrots, turnips, and beets are examples of taproots. The other kind of root system is called a **fibrous root**. These roots are common in grasses and many weeds. These roots hold the plants firmly in place and can extend out from the stem a great distance. Fibrous root plants are often used on sand dunes and beaches to reduce erosion.

Stems

Stems are the major support structure of plants. Remember that plant cells have a cell wall made from **cellulose**, which is a very rigid material. This cell wall gives the plant the strength it needs to stand upright. Plants consist primarily of three types of tissue: dermal, ground, and vascular. **Dermal tissue** covers the outside surface of the stem to prevent excessive water loss and control gas exchange. Ground tissue consists mainly of parenchyma cells and surrounds the vascular tissue providing support and protection. Finally, **vascular tissue** provides long-distance transport of nutrients and water from the soil to the leaves. **Xylem** transports water and minerals, called xylem sap, upward to the leaves. The other vascular tissue is called **phloem** and it carries sugars from the leaves back to the roots for storage. Figure 9.1 shows a generalized plant stem.

Water moves up the stem using a combination of transpiration pull and cohesion tension. **Transpiration** is the loss of water through the stomata. **Cohesion** is the attraction of water molecules to each other. As water evaporates from the leaves, other water molecules are drawn into the roots. To get a better idea of how this works, picture a straw. Take a piece of string and place it into the straw so that some of it is hanging out of the top and some is hanging out of the bottom. Now pull the top of the string. This represents transpiration. As you do this, the string at the bottom moves up (cohesion). Using this process, plants can keep a constant flow of water and nutrients entering into their tissues.

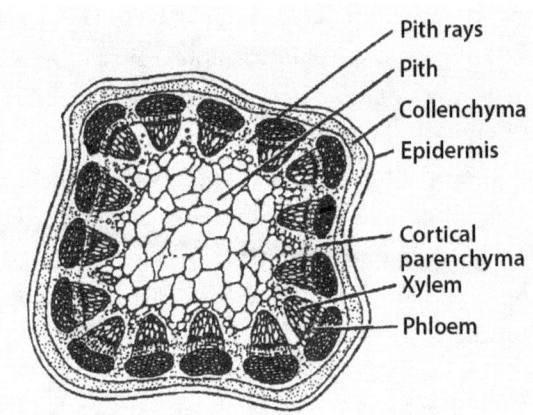

Figure 9.1. Cross Section of Plant Stem

Leaves

One could argue that the leaves are the most important part of a plant. **Leaves** enable plants to capture light and take in carbon dioxide for photosynthesis. Plants exchange gases through their leaves via **stomata**, small openings on the underside of the leaves (Figure 9.2). Stomata allow oxygen to move in or out of the plant and carbon dioxide to move in. Each stoma is protected by guard cells. **Guard cells** are modified cells that regulate when the stomata open and close. These cells behave as a result of water pressure within the plant. When there is a lot of water within the plant tissues (called turgidity), the guard cells swell up, opening the stomata. When turgidity decreases, the guard cells flatten out, closing the opening. This is very beneficial to plants, since they want to reduce water loss when it is in short supply.

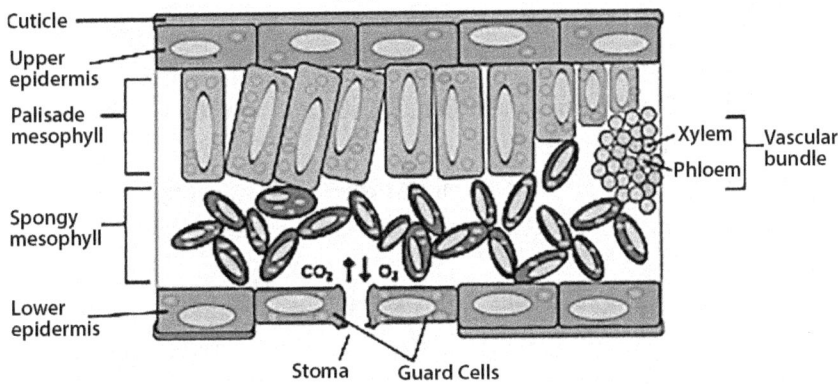

Figure 9.2. Cross Section of Leaf

Photosynthesis occurs primarily in the leaves. The sugar produced by photosynthesis travels down the phloem in phloem sap to the roots and other non-photosynthetic parts of the plant for storage or immediate conversion to energy. You will notice in Figure 9.2 that the leaf actually has a lot of empty space within it. This space is called the **spongy mesophyll**. Some parts of photosynthesis take place here, as well as the exchange of gases (carbon dioxide and oxygen). The denser layer above the spongy mesophyll is called the **palisade mesophyll**. These cells are tightly packed together and are the site of other parts of photosynthesis.

Hormones Drive Plant Behavior

Just as in animals, plants have a variety of hormones that drive their behaviors. Plant hormones, or plant growth regulators, are chemicals secreted internally that regulate growth and development. Plant hormones are present in low concentrations, produced in specific locations, and often act on cells at other locations. The mechanism of hormonal signaling involves attachment of hormone molecules to protein receptors, transmission of the signal along a transduction pathway, and the activation of particular genes.

There are five major classes of plant growth regulators:

- **Auxins** promote cell elongation in growing shoot tips and cell expansion in swelling roots and fruits. In addition, auxins promote apical dominance (the tendency of the main stem of plants to grow more strongly than the side stems) and phototropism (that causes a plant to grow toward light). Auxins stimulate ethylene synthesis (critical in plant germination, ripening, and wound repair).
- **Abscisic acid** (ABA) promotes dormancy, inhibits germination and growth, and responds to water stress by closing stomata.
- **Gibberellins** contribute to stem elongation and stimulate the growth and development of seeds.
- **Ethylene** is the major hormone involved in fruit ripening and abscission (leaf shedding) and induces seed germination, root-hair growth, and flowering. It is a gas.
- **Cytokinins** play a key role in cell division. Cytokinins generally promote shoot development, chlorophyll production, and photosynthesis and inhibit root development.

Tropisms

When a plant grows toward the light (**phototropism**), it is responding to a stimulus. This is called a positive phototropism. In a classic experiment, a researcher placed a bright light on one side of a plant. Within a short time, the plant bent toward the light. The researcher then turned the plant around, facing away from the light. Within a short time, the plant had bent back toward the light.

Other types of responses plants can show are growing in response to gravity (**gravitropism**) and response to touch (**thigmotropism**). Think back to elementary school. You may have performed an experiment using peas or beans where you put them into a wet paper towel and then attached them to a wall. As the shoots started to germinate the roots grew downwards and the stems grew upwards. Now, if you took the shoot and turned it over, the growth would bend in the opposite direction. No matter which way the plant was positioned, roots grow down and stems grow up.

Photoperiods

A photoperiod is the duration of a plant's daily exposure to light. Variations in the length of photoperiods affect its growth, development, and physiological processes. For example, plants adapt to seasonal changes in photoperiods by increasing or decreasing growth processes and changing patterns of photosynthesis and respiration. In fact, there are some plants that are so specialized in their photoperiods that they cannot survive unless they get an exact amount of light. In addition, over time, species of plants develop traits that allow them to thrive in the characteristic photoperiods of their native environments.

Animal Structure and Function

When you are presented with information about digestive systems or reproductive systems, the first thing that comes to mind is how these relate to people. Keep in mind that there are many other animals on the planet that are not human! There are many variations on digestive and reproductive systems, as well as every other cell, tissue, organ, and system in nature. Try not to exhibit "human arrogance" and assume that humans are the dominant species in the universe.

The simplest animals have cells that are differentiated to perform various functions, but they don't have tissues, organs, or organ systems. All animals

have to find and get nutrients, break them down into usable parts, supply energy to cells, excrete wastes, regulate the exchange of gases and chemicals throughout the organism to promote homeostasis, respond to stimuli, and reproduce. What follows is an overview of these main functions in animals with a special focus on humans.

Digestion

The function of the digestive system is to break food down into nutrients and absorb it into the blood stream. From here it can be delivered to all cells of the body for use in cellular respiration. A hydra (which is a microscopic relative of the jellyfish) gets by with a single-opening gastrovascular cavity into which it draws nutrients. The cavity's cells secrete digestive enzymes. Cells with flagella attached to them move the nutrients around and those with pseudopods engulf food particles.

The earthworm has a digestive tract—a tube that runs the length of its body with a mouth that swallows decaying matter and an anus that excretes wastes. Food particles move from the mouth down an esophagus into a crop that stores food, then on to a muscular gizzard that grinds it with particles of ingested sand, and passes on the mixture to the folded walls of the intestines that use chemicals to digest particles and absorb the nutrients. More complex animals show digestion that is an embellishment of this basic plan.

Complex Animal Digestion

When referring to complex animals, it can be said that they have a complete digestive system (both a mouth and an anus) and assorted organs that are responsible for breaking apart the food and absorbing its nutrients. Starting in the mouth, the teeth and saliva begin digestion by breaking down food into smaller pieces and lubricating it so it can be swallowed. Certain enzymes also start working here to help with chemical digestion. The lips, cheeks, and tongue form a **bolus** or ball of food that **peristalsis** (wave-like contractions) carries down the esophagus.

The bolus enters the stomach through the sphincter, which closes to keep food from going back up. In the stomach, hydrochloric acid and other enzymes break down the food, while it is churned by powerful muscles. It is now called **chyme**. The pyloric sphincter muscle opens to allow food to enter the small intestine.

Most nutrient absorption occurs in the small intestine. Its large surface

area, a function of its length and protrusions called **villi** and microvilli, provides the primary absorptive surface into the bloodstream. It is here that accessory digestive organs release additional enzymes that assist with chemical digestion. The gall bladder releases **bile** to help break down lipids and the pancreas releases **trypsin** to help with protein digestion.

Any remaining food then enters the large intestine. The large intestine functions to reabsorb water and produce vitamin K. The feces, or remaining waste, pass out through the anus.

Figure 9.3 shows the human digestive system.

Digestive System

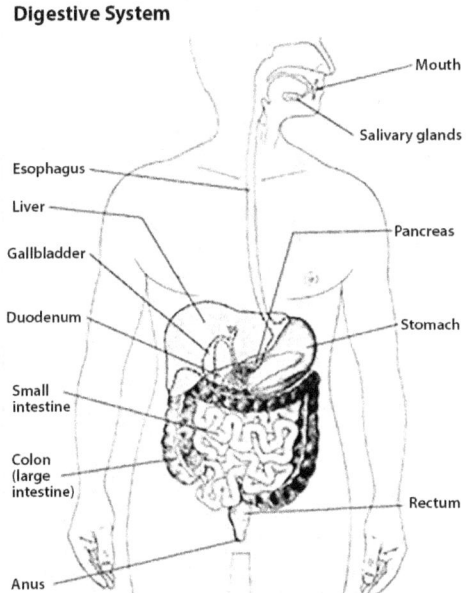

Figure 9.3. Human Digestive System

Circulation

Animals need to be able to circulate nutrients, wastes, and chemicals like enzymes and hormones, throughout their bodies. For organisms like the hydra this presents no problem: their cells are all in contact with the environment so they have no need of a circulatory system. Earthworms, like vertebrates, have a closed circulatory system of tubes—arteries and capillaries—through which a heart pumps oxygenated blood and deoxygenated blood returns through a network of veins.

Complex Animal Circulation

Like the earthworm, more complex animals have a closed circulatory system. This means that oxygenated blood leaves the heart, travels throughout the body, and then returns to the heart to be reoxygenated. The other function of this system is to carry nutrients to all cells of the body and return carbon dioxide waste to the lungs for expulsion. The main organs of the circulatory system include the heart, blood vessels, and blood. Figure 9.4 shows the structure of the human heart (other animals have similar heart structure, although reptiles and amphibians have only three chambers).

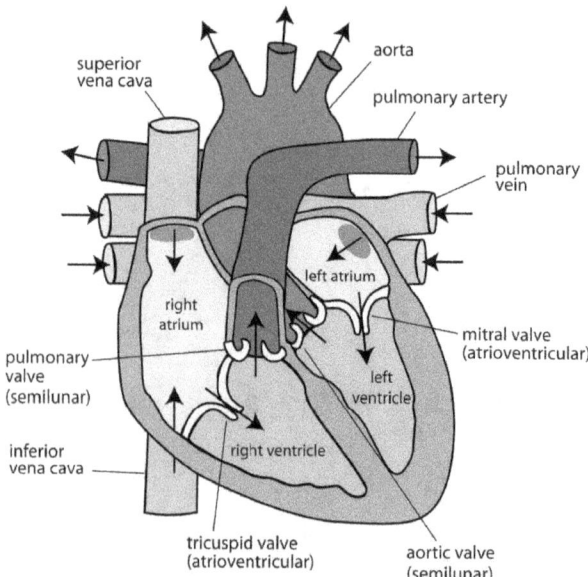

Figure 9.4. The Human Heart

The top two chambers of the human heart are called the **atria**. These chambers receive blood returning to the heart from the body. The blood entering the right atrium is deoxygenated, while that entering the left atrium is oxygenated. The **ventricles** are the lower chambers of the heart. They are responsible for pumping blood out of the heart to either the lungs or the rest of the body.

There are four valves, two atrioventricular (AV) valves and two semilunar valves, inside the heart. The AV valves are located between each atrium and ventricle. The contraction of the ventricles closes the AV valve to keep blood from flowing back into the atria. The semilunar valves are located where the **aorta** (the largest artery in the human body) leaves the left ventricle and where

the pulmonary artery leaves the right ventricle. Ventricular contraction opens the semilunar valves, allowing blood to be pumped out to the arteries, and ventricular relaxation closes the valves.

There are three kinds of blood vessels in the circulatory system: arteries, capillaries, and veins. **Arteries** carry oxygenated blood away from the heart to organs in the body and branch off to form smaller arterioles in organs. Arterioles form tiny **capillaries** that reach every tissue and cell. Downstream, capillaries combine to form larger venules. Venules combine to form larger veins that return blood to the heart. Arteries and veins differ in the direction they carry blood. However, like everything in biology, there is an exception to every rule. Usually arteries carry oxygenated blood away from the heart. The exception to this is the pulmonary artery. Yes, it still carries blood away from the heart to the lungs, but the blood is deoxygenated. Veins usually carry deoxygenated blood to the heart. The exception is the pulmonary vein. Yes, it carries blood to the heart, but in this case, the blood is oxygenated since it is coming directly from the lungs.

Blood is a connective tissue consisting of 60% liquid plasma and several kinds of cells. Plasma contains water salts called electrolytes, nutrients, waste, and proteins. The electrolytes maintain a pH of about 7.4. The proteins contribute to blood viscosity and help maintain pH. Some of the proteins include clotting factors and immunoglobulins, the antibodies that help fend off infection.

The lymphatic system is responsible for returning lost fluid and proteins to the blood. Fluid enters lymph capillaries where it is filtered by lymph nodes that are filled with white blood cells to fight infection.

There are two classes of cells in blood, red blood cells and white blood cells. **Red blood cells (erythrocytes)**—the most numerous—contain **hemoglobin**, which is a protein that carries oxygen. The larger **white blood cells (leukocytes)** are **phagocytic**; they can engulf invaders. White blood cells are not confined to the blood vessels and can enter the interstitial fluid between cells. A third cellular element found in blood is platelets. **Platelets** are made in the bone marrow and assist in blood clotting.

Respiration

Take a deep breath in. Let it out. What did you just do? All animals that need it have developed some method of exchanging oxygen and carbon dioxide with their external environment. The primary structures used to

carry out this exchange are the lungs, but many others have also evolved. Earthworms allow oxygen to diffuse directly through their cell membranes and their skin. Insects and crustaceans have an internal respiratory surface of sinuses. Oxygen enters their bodies through tiny holes called **spiracles**, and passes through a labyrinth of tubes to reach the sinuses. Fish exchange respiratory gases across the complex structure of their gills. Most land vertebrates breathe air into lungs riddled with tiny air sacs called alveoli that are the site of the gas exchange.

Complex Animal Respiration

As the primary respiratory organ of the mammalian respiratory system, the lungs contain a dense network of capillaries just beneath their outer tissue layer. The surface area of this layer is about 100 m^2 in humans. Based upon this surface area, the volume of air inhaled and exhaled in what is called the **tidal volume** is normally about 500 mL in adults. **Vital capacity** is the maximum volume the lungs can inhale and exhale (usually around 3400 mL).

The respiratory system delivers oxygen to the bloodstream and picks up carbon dioxide for release from the body. Air enters the mouth and nose, where it is warmed, moistened, and filtered of dust and particles. Cilia in the **trachea** trap unwanted material in mucus that can be expelled. The trachea splits into two **bronchial tubes**, which divide into smaller and smaller **bronchioles** in the lungs. The internal surface of the lung is composed of **alveoli**, which are thin-walled air sacs lined with capillaries and provide a large surface area for gas exchange. Oxygen diffuses into the bloodstream and carbon dioxide diffuses out of the capillaries to be exhaled out of the lungs. The oxygenated blood is carried to the heart and delivered to all parts of the body by hemoglobin, an iron-rich protein.

The thoracic cavity holds the lungs. The brain controls the rate of breathing. As the rib cage expands and the diaphragm muscle below the lungs drops, the volume of the thoracic cavity increases, and the negative pressure draws in air. When the diaphragm relaxes, we exhale. Since the respiratory system is mainly exposed to the external environment, there are many diseases that can affect it.

Excretion

To poop or not to poop? That is the question. All animals need to rid themselves of metabolic wastes. These may include carbon dioxide and water

(the products of cell respiration) as well as nitrogenous products from protein metabolism. Eliminating nitrogen wastes from the body is essential because a build up of them can be toxic (Yes to the pooping question). Since they live in the water already, marine organisms are able to excrete their nitrogen wastes in a highly concentrated form called ammonia. The water dilutes this into a less hazardous form. Many land animals, however, need to convert ammonia to a less toxic form called urea. Birds and reptiles excrete their urea as a paste called uric acid, while allows them to conserve water. If you have ever seen the results of a flock of birds flying over your freshly washed car, you know about uric acid. Freshwater flatworms have dedicated excretory cells called flame cells that function as single-celled kidneys. Earthworms have paired **nephridia** (modified kidneys), each with a collecting tubule, capillary network, bladder and excretory pore for filtering out nitrogenous wastes. Many arthropods process and excrete wastes in their hindgut through malphighian tubules. Vertebrates use kidneys as the primary excretion organ and produce urea.

Complex Animal Excretion

Both kidneys in the adult human are about 10 cm long. Despite their small size, they receive about 20% of the blood pumped with each heartbeat. The smallest excretory unit of the kidney is the **nephron** shown in 9.5, below.

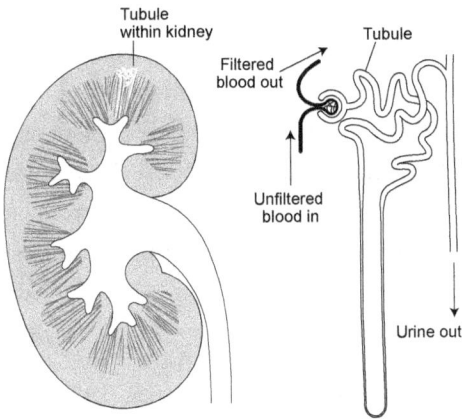

Figure 9.5. Human Kidney

Within each kidney, the Bowman's capsule contains the **glomerulus**, a tightly packed group of capillaries under high pressure in the nephron. The high pressure causes water, urea, salts and other fluids to leak into the Bowman's capsule. This fluid waste (**filtrate**) passes through the three regions of the

nephron: the proximal convoluted tubule, the loop of Henle, and the distal tubule. In the proximal convoluted tubule, waste molecules are secreted into the filtrate. In the loop of Henle, salt is actively pumped out of the tube and much water is lost due to the hyperosmosity of the inner part (medulla) of the kidney. As the fluid enters the distal tubule, water is reabsorbed. Urine forms in the collecting duct that leads to the ureter then to the bladder where it is stored. Urine is passed from the bladder through the urethra. How much water or other fluids an individual has consumed determines both how much water is reabsorbed into the body and how concentrated the urine is.

Structure, Support, and Protection

Maintaining some sort of structure is essential to how an organism survives. Some simple marine and freshwater animals like sponges have a very disorganized structure. They are anchored to the bottom of the ocean and have rigid **spicules** (barbs) that hold the body together. Other marine and freshwater organisms (jellyfish, coral, hydra) as well as worms have internal structures called **hydrostatic** skeletons. These animals use a thick fluid found within their body compartments to hold their shape. Arthropods from crabs to bees have rigid **exoskeletons** made of a polysaccharide called chitin to provide themselves with support, shape, and protection. The downside of having an exoskeleton is that it does not grow along with the animal. When the animal inside of it gets too large, it must shed its skeleton and grow a new one. It is during this time that the animal is very vulnerable. Vertebrates have an internal support structure called an **endoskeleton** made up of bones. Bones grow along with the animal and provide protection for vital internal organs. An exoskeleton limits growth; an endoskeleton does not.

Human skeletons consist of two major divisions. The **axial skeleton** includes the skull and vertebral bones. The **appendicular skeleton** consists of the shoulder girdle, arms, legs and tailbones.

Let's Get Moving

Skeletons not only provide support, shape, and protection, but they also provide structures for muscles to operate and produce movement—to find food, shelter, and mates and to escape predators. Vertebrates have jointed bone limbs that enable movement (arthropods also have jointed limbs). The function of the muscular system is to facilitate movement. There are three types of vertebrate muscle tissue: skeletal, cardiac, and smooth.

Skeletal muscle is voluntary (subject to conscious control). These muscles are attached to bones and are responsible for their movement. Skeletal muscle consists of long fibers and is striated due to repeating patterns of myofilaments (made of the proteins **actin** and **myosin**) that make up the fibers.

Cardiac muscle, found only in the heart, is striated like skeletal muscle but differs in that the plasma membrane of the cardiac muscle causes its cells to beat. Cardiac muscle is involuntary (you do not have to think about making your heart beat as you would raising your arm over your head).

Smooth muscle is also involuntary (not subject to conscious control). It is found in organs like the lungs and intestines and enables functions such as digestion and respiration. Unlike skeletal and cardiac muscle, smooth muscle is not striated. Smooth muscle has less myosin and does not generate as much tension as the striated muscles.

Skeletal muscle contracts when a nerve impulse strikes a muscle fiber, which causes calcium ions to flood the sarcomere. The myosin fibers creep along the actin, causing the muscle to contract. Once the nerve impulse has passed, calcium is pumped out and the contraction ends. This is called the **sliding filament theory**. See figure 9.6 below.

Figure 9.6. Sliding Filament Theory

Information Transmission

It is essential that living things be able to sense stimuli in their surroundings and respond to them accordingly. There are many different ways they accomplish this, but the most common is using some form of nerve cell. As organisms become more complex, the concentration of nerve cells increases. Sponges have no nerves at all. Earthworms and insects have very simple nervous systems, often containing fused **ganglia** (groups of nerves) that connect to a nerve cord. One adaptation that led to the development of concentration of nerve cells is **cephalization**, which means the formation of a head region. Nerve cells are often concentrated in this region, providing extra sensory capabilities. In vertebrates, there are highly sophisticated nervous systems that enable the animals to respond to external stimuli and also learn from their various experiences.

Complex Organism Signal Transfer

The **neuron** (Figure 9.7) is the basic unit of the nervous system. It consists of an axon, which carries impulses away from the cell body; the dendrite, which carries impulses toward the cell body; and the cell body, which contains the nucleus. The myelin sheath, composed of Schwann cells, covers the neuron and provides insulation, which allow electrical impulses to travel quickly through the neuron. Synapses are junctions between neurons. Chemicals called neurotransmitters serve as signaling molecules that are released from one neuron and diffuse through the synaptic cleft to another neuron.

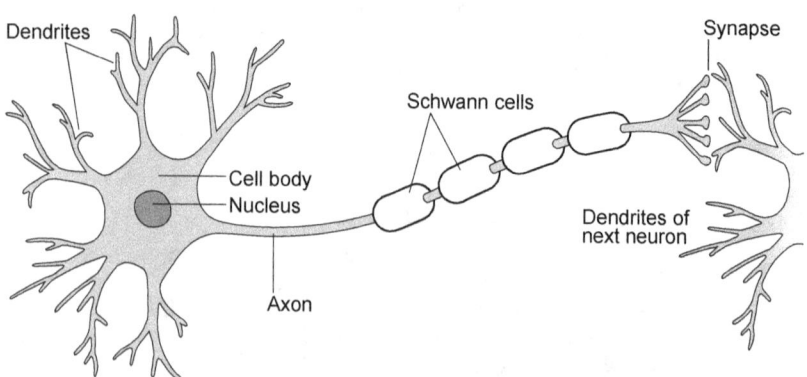

Figure 9.7. Nerve Cell (Neuron)

Nerve action depends upon an imbalance of electrical charges between the inside of the neuron and the outside. These electrical charges are carried

by ions such as sodium, calcium, and potassium. When the ions move from one side of the neuronal membrane to the other (from outside the cell to inside or vice versa) an electrical current flows through the neuron. These electrical currents are called **action potentials**. Action potentials trigger the release of neurotransmitters from the axon into the synaptic cleft. When the neurotransmitters diffuse through the synaptic cleft, they bind to receptors on the surface of dendrites. This binding then triggers another action potential in the next neuron.

When a neuron is resting, it has a negative charge and is said to be hyperpolarized. When ions like sodium flow into the neuron, it takes on a positive charge, becomes depolarized, and an action potential is generated.

Divisions of the Nervous System

In most animals the nervous system is divided into two main parts. The **central nervous system (CNS)** consists of the brain and spinal cord. The CNS is responsible for the body's response to environmental stimuli. The spinal cord sends out motor commands in response to stimuli that are automated or reflexive. The brain is where responses to more complex stimuli occur. The **meninges** are the connective tissues that protect the CNS. The CNS contains fluid-filled spaces called ventricles that are filled with cerebrospinal fluid that cushions the brain and spinal cord and circulates nutrients, white blood cells, and hormones.

The **peripheral nervous system (PNS)** consists of the nerves that connect the CNS to the rest of the body. The sensory division brings information to the CNS from sensory receptors, and the motor division sends signals from the CNS to effector cells. The motor division consists of the **somatic nervous system** and the **autonomic nervous system**. The autonomic nervous system is controlled unconsciously by the hypothalamus of the brain and regulates the body's internal environment. This system is responsible for the movement of smooth and cardiac muscles as well as the muscles for other organ systems. The somatic nervous system is controlled consciously in response to external stimuli.

Endocrine System

You have probably heard the expression that teenagers are nothing more than a pile of raging hormones. Well, this is because during puberty the body releases a variety of chemical messengers that cause a lot of physical and

emotional changes. Responses like this are not limited to just people. Many organisms (plants included. Remember auxin and gibberellin?) produce hormones that induce changes within them.

Many hormones are responsible for the development of reproductive organs and preparation for mating. However, there are several others that perform other tasks to help an organism maintain homeostasis. What is homeostasis, you ask? Well, it is an organism's ability to maintain a constant internal environment. Think about the last time you were hot, maybe after running or playing tennis in gym class. Your internal body temperature had increased due to all the exercise you performed. To attempt to bring that temperature back down to normal, your body will start to perspire. As the sweat evaporates from your skin, it carries heat away with it. This cools you down.

Situations like this, where the body is trying to return to a normal condition, are called **negative feedback**. Another example would be shivering when you are cold.

As you have guessed, with all things negative, there also has to be a positive. **Positive feedback** is the way the body increases an existing condition. While there are times when positive feedback is a good thing, most of the time it takes the body away from homeostasis. For example, if you suffer a serious wound and are bleeding, the loss of blood will cause a drop in blood pressure. The brain recognizes this and sends a signal to the heart to pump faster. This increases the amount of blood flowing. Since you are wounded, this will cause more blood loss.

Hormones In Complex Organisms

The function of the complex organism **endocrine system** is to manufacture proteins called hormones that help regulate the body, often in concert with the nervous system. **Hormones** are released into the bloodstream and carried to a target tissue where they stimulate an action. Hormones are specific and fit receptors on the cell surface of the target tissue. There are two classes of hormones: steroid and peptide. Steroid hormones come from cholesterol and include the sex hormones testosterone and estrogen. Peptide hormones are derived from amino acids.

Hormones are secreted by endocrine cells, which make up endocrine glands. The major endocrine glands and their hormones are as follows:

- **Hypothalamus** – located in the lower brain; signals the pituitary gland

- **Pituitary gland** – located at the base of the hypothalamus; releases growth hormones and antidiuretic hormone (causing retention of water in kidneys)
- **Thyroid gland** – located on the trachea; lowers blood calcium levels (calcitonin) and maintains metabolic processes such as heart rate, blood pressure, muscle tone, digestion, and reproductive functions (thyroxine)
- **Parathyroid glands** – located next to thyroid gland: maintain the calcium level in blood
- **Gonads** – located in the testes of the male and the ovaries of the female; testes release androgens to support sperm formation and ovaries release estrogens to stimulate uterine lining growth and progesterone to promote uterine lining growth, regulate the development of the male and female reproductive organs
- **Pancreas** – located in the abdomen; secretes insulin to lower blood glucose levels and glucagon to raise blood glucose levels

The functions of the nervous and endocrine systems are closely entwined. Neurotransmitters are chemical messengers. The most common neurotransmitter is acetylcholine. Acetylcholine controls muscle contraction and heartbeat. A group of neurotransmitters, the catecholamines that include epinephrine (adrenaline) and norepinephrine, are also hormones and are produced in response to stress. These hormones/neurotransmitters increase the rate and stroke volume of the heart, thus increasing the rate of oxygen delivery to the blood cells and thus have profound effects on the cardiovascular and respiratory systems.

The Immune System

Throughout the living world, organisms have evolved ways of defending themselves against invading parasites and pathogens. Some of the chief methods are phyagocytosis (engulfing and destroying) the invader, encapsulating it, or poisoning it with chemicals. Plants and fungi have evolved many chemical defenses against parasites and predators. These chemicals often make the organism toxic to the invader.

Other organisms of interest for their immune systems are the horseshow crab and different species of sharks. Both these animals have been on Earth for millions of years and do not seem to be susceptible to diseases. The blood of

horseshoe crabs is unusual because in addition to being blue, it is also based on copper. Sharks seem to be immune to cancer and most diseases due to anomalies with their white blood cells. Both of these animals are being studied extensively to see if these powers can be used to improve the human immune system.

Defense in Complex Organisms

The immune system in complex organisms has two types of defense mechanisms against foreign invaders: non-specific and specific.

The **non-specific** immune mechanism is composed of two parts. The body's physical barriers—the skin and mucous membranes—are the first line of defense. As long as there are no abrasions on the skin, that can prevent the penetration of bacteria and viruses. Mucous membranes form a protective barrier around the digestive, respiratory, and genital tracts (remember you read that the respiratory system was totally exposed to the external environment so it was most likely to become infected? Well, these mucus membranes evolved to help protect it). In addition, the pH of the skin and mucous membranes inhibit the growth of many microbes. Mucous secretions (tears and saliva) wash away many microbes and contain lysozymes that kill many microbes.

The second component of the non-specific immune response includes white blood cells and inflammatory responses. Some white blood cells (neutrophils, eosinophils, and monocytes) engage in **phagocytosis**, the process of using the cell membrane to engulf foreign particles and form an internal (neutralized) phagosome. Neutrophils make up about 70% of all white blood cells. Monocytes mature to become macrophages, which are the largest phagocytic cells.

Instead of killing the invading microbe directly, natural killer cells destroy the body's own infected cells. During an inflammatory response, blood supply to the injured area is increased, causing redness and heat. Swelling also typically occurs with inflammation. Histamine is released by basophils and mast cells when cells are injured triggering the inflammatory response.

The **specific** immune mechanism recognizes specific foreign material (individual pathogens) and responds by destroying the invader. An **antigen** is any foreign particle that elicits an immune response. The body manufactures an **antibody** to recognize and latch onto antigens and destroy them. Memory of the invaders provides immunity upon further exposure. This is why most people only get the mumps one time.

Immunity is the body's ability to recognize and destroy an antigen before it causes harm. Active immunity develops after recovery from an infectious disease (e.g., chickenpox) or after a vaccination (e.g., mumps, measles, and rubella). Passive immunity may be passed from one individual to another and is not permanent. A good example is the immunity passed from mother to her nursing child. A baby's immune system is not well developed and the passive immunity he or she receives through nursing provides additional protection.

After exposure to an antigen, the body produces either a humoral or a cell-mediated response.

1. **Humoral response:** Free antigens and antigen-presenting cells activate B cells (lymphocytes from bone marrow) that transform into plasma cells that secrete antibodies. Memory cells are also generated that recognize future exposure to the same antigen. Antibodies defend the body against extracellular pathogens by binding to the antigen and making it an easy target for phagocytes to engulf and destroy. Antibodies are in a class of proteins called immunoglobulins.
2. **Cell-mediated response:** Infected cells activate T cells (lymphocytes from the thymus) that then bind to the infected cells and destroy infected host cells along with the antigen.

Vaccines are antigens given in very small amounts that stimulate both humoral and cell-mediated responses and help memory cells recognize future exposure to the antigen so the body can produce antibodies much faster. When you get a flu shot, the doctor is injecting dead forms of all the known flu viruses into your body. This initiates a response of the immune system to produce antibodies against all of them. If you then come in contact with any of the forms of the virus, your body has already seen it, so the chances of fighting it off sooner are increased.

The immune system attacks not only microbes, but also cells that are not native to the host. Skin grafts, organ transplantations, and blood transfusions are all examples of things that your body can defend against. Antibodies to foreign blood and tissue types already exist in the body. If blood is transfused that is not compatible with the host, these antibodies destroy the new blood cells. There is a similar reaction when tissue and organs are transplanted.

The major histocompatibility complex (MHC) is responsible for the rejection of tissue and organ transplants. This complex is unique to each person.

Since cytotoxic T cells recognize the MHC on transplanted tissue or organ as foreign and destroy these tissues, a transplant recipient needs various drugs to suppress the immune system and prevent rejection of foreign tissue, but this also leaves the patient more susceptible to infection. Autoimmune disease occurs when the body's own immune system destroys its cells. Lupus, Grave's disease, and rheumatoid arthritis are examples of autoimmune diseases. There is no way to prevent autoimmune diseases. Immunodeficiency (for instance, HIV) is a deficiency in either the humoral or cell-mediated immune defenses.

Reproduction and Development

The main drive living things have is to reproduce. Everything they do, whether it be eating, finding shelter, or migrating, helps to ensure that their genes will get passed on. Many different reproductive mechanisms have evolved. Bacteria simply duplicate their DNA and then divide in half. Plants, as you will read below, use pollen and seeds to pass on their genes. Flowering plants in particular have very elaborate reproductive behaviors that involve wind and weather conditions, as well as animals. In fact, some plants cannot start to reproduce until their environment has been burned to the ground. Animal reproductive behavior is also quite varied and remarkable. Many birds do dances to attract their mates, while male elephant seals battle almost to the death for the right to mate with females. However they do it, living things need to be sure they get their genes passed on.

Reproduction and Development in Non-vascular Plants

These seedless plants (mosses, liverworts, and hornworts) can reproduce asexually (quick and less costly, especially suited for unstable environments) or sexually with spores. In asexual reproduction, a fragment of a plant, typically a leaf, or a more-or-less developed frond breaks off and grows into a new individual. In the sexual stage, the plant produces **haploid** spores. The sperm cells need water to swim to the eggs. As a result, these plants live in environments that tend to stay moist all the time. The fertilized zygote, called a **sporophyte** develops into a sexual adult, and the cycle repeats.

Reproduction and Development in Vascular Plants

Alternation of generations

Reproduction in vascular plants is accomplished through **alternation of generations** (Figure 9.8). Simply stated, a haploid stage in the plant's

life history alternates with a diploid stage. The diploid sporophyte divides by meiosis to reduce the chromosome number to the haploid gametophyte generation. The haploid gametophytes undergo mitosis to produce gametes (sperm and eggs). Finally, the haploid gametes fertilize to return to the diploid sporophyte stage.

Like the non-vascular plants, the vascular, non-seeded plants (including horsetails and ferns) also reproduce with spores that need water to enable the sperm to swim to the egg. Gymnosperms and angiosperms use seeds for reproduction and do not require water.

Angiosperms generally reproduce sexually though some can reproduce asexually as well. Angiosperm sexual reproductive structures are called flowers. Most flowers are called **complete**, having both male and female reproductive parts. **Incomplete** flowers have either male or female reproductive parts. Figure 9.9 shows a complete flower.

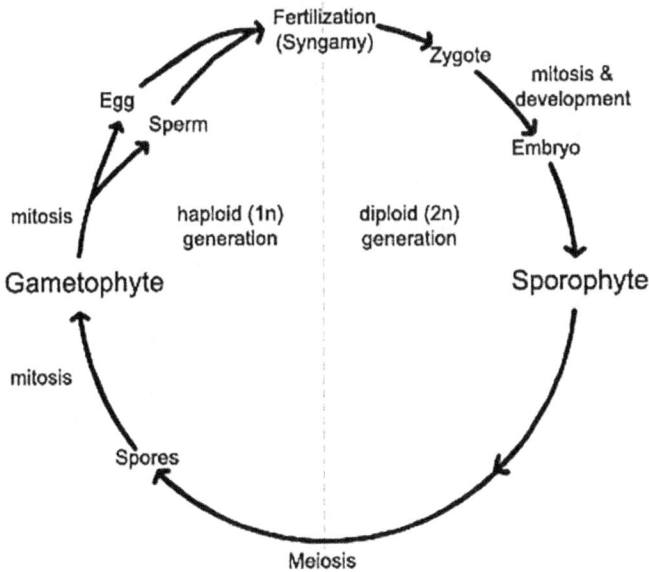

Figure 9.8. Alternation of Generations

The male part of a flower is called the **stamen**. It is composed of the **anther**, where pollen develops, and the **filament**, a stalk-like structure that holds up the anther. The female reproductive structure is called the **carpel**. It consists of the **stigma** (a sticky landing pad that collects the pollen), the style (a tube that connects the stigma to the ovary), and the **ovary** (where the ovule develops into a seed).

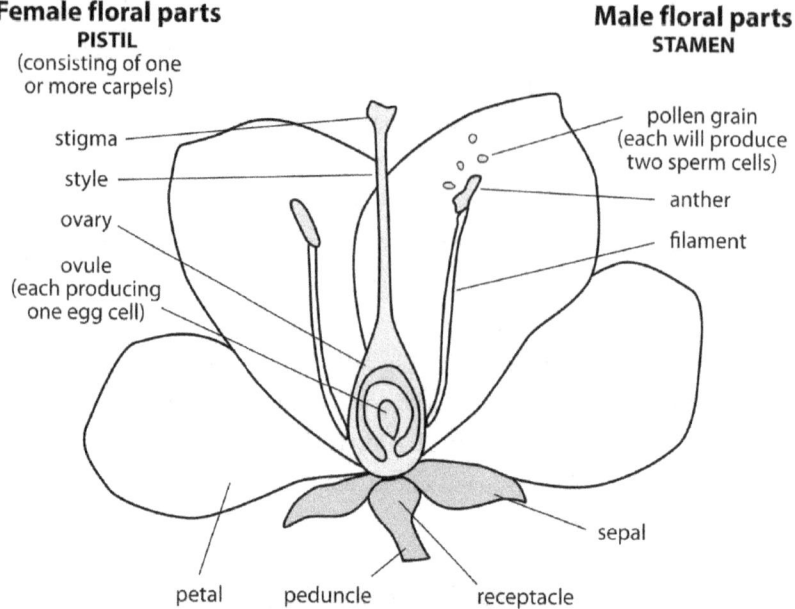

Figure 9.9. Anatomy of a Flower

During **pollination,** plant anthers release pollen grains, which can be carried by animals or the wind to plant carpels. During the late spring and early summertime in the northern climates, it is very easy to see when pollen is being released. It often covers sidewalks and automobiles with a yellow dust. Some plants self-fertilize, so the pollen falls onto the carpels of the same flower.

The sperm is released to fertilize the eggs. Angiosperms reproduce through a method called **double fertilization**. Two sperm fertilize a single ovum. One sperm produces the new plant, and the other forms the food supply for the developing plant (**endosperm**). The ovule develops into a seed, and the ovary develops into a fruit. In many cases, to avoid competition with the parent plant, the wind or animals carry the seeds to new locations, and new plants form in a process called **dispersal.**

Germination describes the initiation of growth of an organism from the seed. When conditions are favorable for angiosperms, a seedling sprouts or germinates from a seed that contains the stored embryo and food reserves. Once this happens, the plant embryo resumes growth and continues to develop into a mature plant. In addition, cellular plant hormones (like auxins and ethylene) direct the germination and growth processes.

Vegetative propagation refers to the ability of some types of plants to reproduce asexually by generating new plants from existing vegetative structures. Examples of vegetative propagation include the formation of new plants from long underground stems, root sprouting, and budding from leaf edges. Natural vegetative propagation is most common in perennial plants (plants that come back every year). Vegetative propagation is a form of cloning, because the offspring plants are identical to the parent. Humans often take advantage of the regenerative nature of plants to simplify plant breeding and propagation. Examples of human-made vegetative propagation are cuttings and graftings used in orchards to produce specific, desirable varieties of fruit.

Reproduction and Development in Animals

Like plants, other organisms can reproduce sexually or asexually. Bacteria are the champion asexual reproducers. Most animals reproduce sexually, but there are some exceptions to that (remember the saying here is that everything in biology happens for a reason and there is an exception to every rule).

Asexual Reproduction Strategies

Asexual reproduction does not require the union of egg and sperm to create a new organism. It offers advantages to individuals that don't have easy access to mates but live in a stable environment where genetic variation matters less because offspring face fewer challenges requiring adaptation. Asexual reproduction is quick and energetically "cheap," because it does not require a reproductive system or hormones, the generation of gametes, or any of the costly behaviors involved in finding and competing for a mate. There are three common types of asexual reproduction seen in more complex organism.

A simple organism like a hydra can reproduce by **budding**. This means generating buds on the tips of its tentacles that separate from the parent and become cloned offspring.

Sponges, sea stars, and flatworms reproduce by **fragmentation**. Here, the adult literally goes to pieces, regenerating missing body parts. If you picture a sea star, you know that they usually have a central disk from where the arms radiate. If you cut a sea star into pieces, but each still has a portion of the central disk, an entirely new sea star can regenerate.

Parthenogenesis occurs when an embryo grows and develops without fertilization from a female with no contribution from a male. Parthenogenesis does not simply produce clones. It is a true reproductive process that creates new individuals from the varied genetic material of the mother. Clearly, however, with all the genetic material coming from a single parent, there will be less overall genetic variability. All the offspring of parthenogenetic embryos are female, but the offspring themselves may be capable of sexual reproduction, parthenogenesis, or no reproduction at all. Parthenogenesis occurs naturally in invertebrates including water fleas, aphids, and honeybees and in vertebrates such as lizards, salamanders, and certain fish.

Certain species, such as the whiptail lizard reproduce exclusively via parthenogenesis. Other species may alternate between parthenogenesis and sexual reproduction. Certain environmental stimuli may trigger parthenogenesis. For example, aphids use parthenogenesis when there is ample food because parthenogenesis is quicker (hence more efficient) than sexual reproduction.

Sexual Reproduction Strategies

Like the vascular plants, most animals go through life as diploid individuals that produce haploid gametes through meiosis. Even many jellyfish, flatworms, and roundworms that can reproduce asexually reproduce sexually as well. The major advantage that sexual reproduction confers on a species is genetic variation that can help it survive in an unstable environment.

For the most part, sexual reproduction requires that motile sperm fertilize sessile eggs and produce a diploid zygote that develops into an embryo. Remember that meiosis produces these haploid gametes. While in vertebrates the sexes are generally well defined, among invertebrates there has been more evolutionary experimentation. For instance, many snails, barnacles, and earthworms are **hermaphrodites**, harboring both male and female reproductive systems in one individual and may fulfill both male and female roles when mating.

Development

Development is defined as a change in form. The first divisions of the fertilized egg are called **cleavage** that continues until the egg becomes a **morula**, which is a solid ball of cells, which continues to divide until it becomes a **blastula**, a hollow ball of undifferentiated cells. Further cell

division yields a structure called a **gastrula**. **Gastrulation** defines the start of tissue differentiation into the separate germ layers, the endoderm, mesoderm, and ectoderm (depending on the species). Each of these germ layers will give rise to the internal and external organs of the organism. **Neuralation** marks the development of the nervous system while **organogenesis** denotes the development of the various organs of the body.

Reproduction and Development in Complex Organisms

Most events that happen in complex organism reproduction involve hormones. Maturation, puberty onset in humans, coming into estrus, and even pheromone production are all regulated by various hormones (see the endocrine review already discussed).

Hormones regulate reproduction. In males, the primary sex hormones are the androgens, **testosterone** being the most important. The androgens are produced in the testes and are responsible for the primary and secondary sex characteristics of the male. Female hormone patterns are cyclic and complex. Most women have a reproductive cycle, called the menstrual cycle, of about 28 days.

The **menstrual cycle** is specific to the changes in the uterus. The ovarian cycle that results in ovulation is regulated by hormones and occurs in parallel with the menstrual cycle. Five hormones participate in this regulation, most notably estrogen and progesterone, which play an important role in signaling the uterus and in development and maintenance of the endometrium. Estrogens are also responsible for secondary sex characteristics of females.

Sperm and egg cells are produced during the process of **gametogenesis**. You remember from the review of meiosis that male mammals have an indefinite reproductive life. **Spermatogenesis** begins at puberty in the male. One spermatogonia, the diploid precursor of sperm, produces four sperm. Immature sperm are produced in the seminiferous tubules located in the testes. After leaving the seminiferous tubules, sperm mature and are stored in the epididymis located on top of the testes. After ejaculation, the sperm travel up the **vas deferens** where they mix with semen made in the prostate and seminal vesicles and travel out the urethra.

Oogenesis, the production of egg cells (ova), is usually complete by birth in a female. Remember that female mammals often have a finite reproductive life. They are born with all of the eggs they are ever going to

have. Egg cells are not released until menstruation begins at puberty. Meiosis forms one ovum with all the cytoplasm and genetic material and three polar bodies that are reabsorbed by the body. The ova are stored in the ovaries and released each month from puberty to menopause.

Ovulation releases the egg into the **fallopian tubes**. These tubes are lined with cilia that move the egg along. Fertilization of the egg by the sperm normally occurs in the fallopian tubes. If pregnancy does not occur, the egg passes through the uterus and is expelled through the vagina during menstruation. Levels of progesterone and estrogen stimulate menstruation and are affected by the implantation of a fertilized egg.

If fertilization does occur, the **zygote** (the fertilized egg) begins dividing in approximately 24 hours. The resulting cells form a blastocyst that implants two to three days later into the wall of the uterus. Implantation promotes secretion of human chorionic gonadotrophin (HCG), which keeps the level of progesterone elevated to maintain the uterine lining in order to feed the developing embryo until the umbilical cord forms. Pregnancy tests detect HCG.

Organogenesis, the development of the body organs, occurs during the first trimester of fetal development. The heart begins to beat and all the major structures are present at this time. The fetus grows very rapidly during the second trimester of pregnancy. The fetus is about 30 cm long and is very active during this stage. During the third and last trimester, fetal activity may decrease as the fetus grows. The hormone oxytocin initiates labor, which causes dilation of the cervix and labor contractions. Prolactin and oxytocin cause the production of milk.

Animal Behavior

Have you ever seen baby ducks walking in a line behind their mother? How about a cat running into the kitchen whenever it hears the can opener? If so, then you have seen some of the diversity of animal behaviors that exist. Animal behavior is responsible for courtship leading to mating, communication between species, territoriality, aggression between animals, and dominance within a group.

Behaviors can be **innate**, meaning the animal is born with them, or **learned**, which means they develop them through their experiences. Behaviors are usually carried out as the result a stimulus. Something occurs

that causes the animal to need to respond to its environment. For example, when the temperature drops and the days get shorter, bears know that it is time to start their hibernation period.

The ducks mentioned above are an example of a behavior called **imprinting**. Many animals exhibit a strong mother-child bond. Since the mother is usually the first thing the offspring see after hatching, the baby ducks immediately learn that this animal will protect them and take care of them. As a result, they follow her around everywhere. As the ducklings age and no longer need their mother, the effects of the imprinting wear off. Many studies have been performed to see if ducklings and other waterfowl would imprint on other things besides their mother. These experiments were successful, getting the ducklings to imprint on the researchers.

Learned behaviors can fall into one of two categories. **Classical conditioning** occurs when an animal associates some form of stimulus with a natural behavior. The legendary work of Ivan Pavlov showed that dogs could be conditioned to salivate at the sound of a bell. Whenever dogs are presented with a piece of meat they start to drool. Pavlov took this natural behavior and added the sound of a bell whenever he gave the dogs meat. Over time, the dogs would drool whenever they heard the bell, even if there was no meat present. The cat's reaction to the sound of the can opener above is another example of classical conditioning.

Operant conditioning happens through trial and error. This form of learning depends on the life experiences of the animal. Say a mouse is placed into a cage with a running wheel. After running in the wheel, some food automatically falls out of a trapdoor. After enough exposure to this, the mouse will figure out that it needs to run in order to get the food.

Animal Social Behavior

Usually, animals are very social organisms. They often live in groups, or at least come together for mating purposes. When they do, they need to communicate with each other. They use body language, sound, and smell to relay messages between themselves. Perhaps the most common type of animal communication is the presentation or movement of distinctive body parts. Many species of animals reveal or conceal body parts to communicate with potential mates, predators, and prey. In addition, many species of animals communicate with sound. Examples of vocal communication include the

mating "songs" of birds and frogs and warning cries of monkeys. Finally, many animals release scented chemicals to communicate with other animals. These chemicals are called **pheromones** and are important in reproduction and mating.

Chapter 10: Ecology

Life on Earth is an immensely complex, constantly shifting, ever-evolving drama that involves organisms interacting with each other and with their environment. Studying those relationships and what drives them is the field of ecology. All life depends upon energy from the sun, so the basis of ecology is understanding how energy and nutrients flow through systems. Examining species populations, communities of organisms, ecosystems, and biomes shows how the web of life sustains itself and what perils it faces. Humans are having a dramatic impact on the planet by causing things like pollution and climate change. We are also challenging the overall sustainability and carrying capacity of the planet.

Energy Gets Its Groove On

One of the primary things the field of ecology does is examine how energy gets transferred through an ecosystem. Remember that energy can neither be created nor destroyed, just converted. Feeding relationships between organisms determine energy flow and chemical cycling within a food web and an organism's position (its **trophic level**) in it. **Autotrophs** are the primary producers of the ecosystem—mainly plants and green algae – and the basis of most energy flow. An autotroph ("self-feeder") is an organism that makes its own food from the energy of the sun or other elements. Autotrophs include

- photoautotrophs that make food from light and carbon dioxide, releasing oxygen that can be used for respiration.
- chemoautotrophs that oxidize sulfur and ammonia in the absence of oxygen. Some bacteria are chemoautotrophs.

All other organisms are **heterotrophs** ("other feeders") organisms that must eat other living things to obtain energy. Another term for heterotrophs is **consumers** and can either be an **herbivore** (plant eater) or a **carnivore** (meat eater). Sometimes animals fall into both categories and eat both plant matter and meat. Bears are often considered **omnivores** because they eat both.

All animals are heterotrophs. The **primary consumers** are the herbivores that eat plants or algae (the primary producers). **Secondary consumers** are the carnivores that eat the primary consumers. **Tertiary consumers** eat the secondary consumers. These trophic levels may go higher depending on how much energy is available. Most chains will not have more than five levels because the amount of energy available to those organisms at the very top is too limited to support them.

Scavengers like vultures eat dead things. They are still considered consumers but do not fall into a particular trophic level.

All trophic levels are consumed by decomposers. **Decomposers**, such as bacteria and fungi, are consumers that break down animal waste and dead organisms. They release the nutrients found within the dead organism's bodies back into the environment. The carbon, nitrogen, and phosphorus they release serve as the basis for new life.

This pathway of food transfer is depicted by a food chain. Since many animals eat more than one thing they get their energy from a variety of different sources. Therefore, a simple chain is not representative of how the energy is actually transferred. Figure 10.1 shows a food web, which is a more realistic model of an ecosystem's energy transfer.

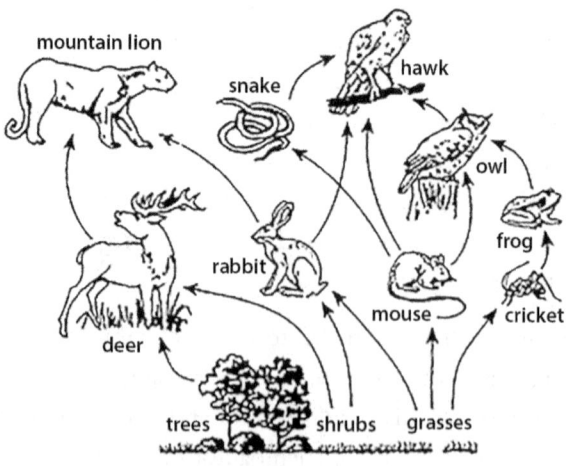

Figure 10.1. Food Web

In a food web, the arrows always point to what does the eating. In the figure above, the mouse eats the grasses, the owl eats the mouse, and the hawk eats the owl.

Biological Magnification

You know that food chains rarely have more than five trophic levels because the amount of energy available to the top consumers is limited. Another thing that happens in food chains is that the higher up the food chain an organism is, the more toxins it may accumulate, a process called **biological magnification**. For example, red tides are caused by microscopic organisms called dinoflagellates. These dinoflagellates produce toxins and are a primary food source for clams and mussels, which filter the dinoflagellates out of the water as they are feeding. The toxins get incorporated into the tissue of the clam or mussel, but not in large enough amounts to harm it. When a larger animal, such as a sea otter, eats the clams, however, it can be poisoned because it needs to eat many clams to survive, and the toxins become concentrated in its body.

Round and Round They Go – Nutrient Cycles

Biogeochemical cycles are nutrient cycles that involve both biotic and abiotic factors. Primary consumers and decomposers are especially critical in cycling the building blocks of ATP and cell nutrients from soil, water, and air into organisms and back to the earth again. There are several nutrient cycles that are important to living things.

Water cycle: Two percent of all the available (fresh) water on Earth is fixed and unavailable in ice or the bodies of organisms. The water that is available can be found in surface water (e.g., lakes, oceans, rivers) and ground water (e.g., aquifers, wells). Ninety six percent (96%) of all available water is ground water. Solar energy drives the water cycle. The Sun heats the water on Earth, causing it to evaporate. In the air, which is cooler, the water vapor condenses and falls back to Earth as precipitation. It then runs over the land and pools in the oceans, lakes, streams, and ponds. Here, it heats up again.

Plants also lose a lot of water through transpiration. Water is pulled up through the roots into the stems and leaves, then is lost through the leaves into the atmosphere. Figure 10.2 shows a model of the water cycle.

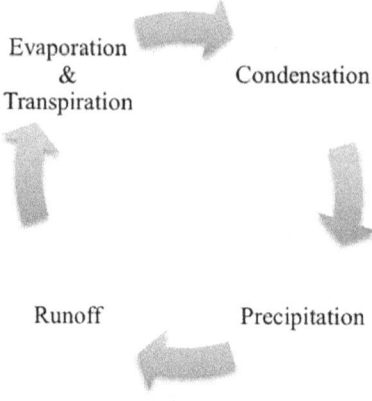

Figure 10.2. The Water Cycle

Carbon cycle: Ten percent (10%) of all available carbon in the air (in the form of carbon dioxide gas) is fixed by photosynthesis. Plants fix carbon in the form of glucose. Animals eat the plants and are able to obtain carbon and energy through cellular respiration. When animals release carbon dioxide through respiration, the plants again have a source of carbon for further fixation. The burning of fossil fuels also contributes to the amount of carbon in the atmosphere. Cars and emissions from factories and coal-fired power plants are contributing both to the increased amount of carbon dioxide and temperatures on Earth.

Nitrogen cycle: Eighty percent (80%) of the atmosphere exists in the form of nitrogen gas. Most nitrogen is useless to living things in this form. It must first be fixed and taken out of the gaseous form so that living things can use it. Many plants have symbiotic bacteria living in their roots that pull the nitrogen out of the atmosphere and convert it into a useable form. Called **nitrogen fixation**, peas, beans, and alfalfa are examples of plants that have these symbiotic bacteria. Nitrogen is necessary to make amino acids and the nitrogenous bases of DNA.

Populations

A **population** is a group of individuals of one species that live in the same general area at the same time. Ecologists are always concerned with the size of different populations, especially if the organism in question is endangered. The four main factors on which population size depends include births, deaths, immigration (entering the area), and emigration (leaving the area). The total number of individuals the habitat can support also plays a role in population size. This factor is called the **carrying capacity** and is determined by the amount of

resources available. Most of the time a population curve (a visual representation of its size over time) starts with rapid growth. This happens because there are plentiful resources for all new members. Over time, the population gets so large that the once plentiful resources become limited. At this point, the population has reached the maximum number of individuals it can support. As long as the resources maintain themselves, then the population size will remain fairly constant (see Figure 10.3). The fluctuations seen in the graph are natural occurrences of variation due to "good years" and "bad years."

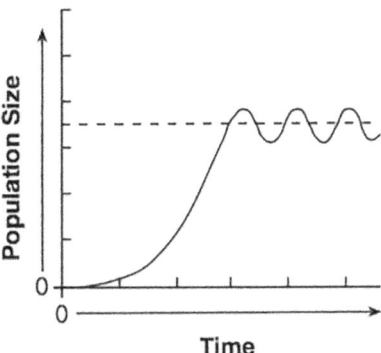

Figure 10.3. Carrying Capacity

The individuals of a population are always fighting for food, mates, shelter, and other resources. **Competition** occurs when resources are limited and more than one individual or population in a community needs them. As a result of competition, population sizes can be affected. Since one individual or population is often better at getting the needed resources, the other has to either adapt to use other resources, perish, or leave.

Limiting factors also affect population growth. As a population size increases, competition for resources intensifies, and the growth rate declines. This is called a **density-dependent** growth factor. **Density-independent factors** affect individuals regardless of population size. Weather and climate are good examples. Temperatures that are too hot or cold can kill many individuals, causing a decrease in the population size.

Exponential growth occurs when there is an abundance of resources and the growth rate is at its maximum, called the **intrinsic rate of increase**. This relationship can be graphically represented in a growth curve (see Figure 10.4). An exponentially growing population begins with little change and then rapidly increases.

Many populations follow this model of population growth. Humans are an exponentially growing population. Eventually, we will reach the carrying capacity of the Earth, and the growth rate will level off.

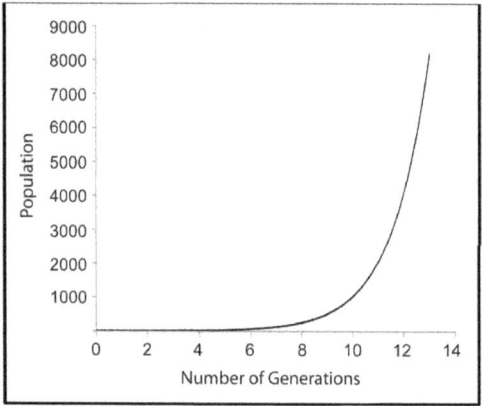

Figure 10.4. Exponential Growth Curve

Population density is the number of individuals per unit area or volume. The spacing pattern of individuals in an area is called **dispersion**. Clumped dispersion is the most common and realistic form of resource allotment. Most resources are found together in clumps, but there is much space between each of them. Uniform dispersion patterns suggest a non-random placement of resources, such as corn growing in a field. Each corn stalk is equidistant from every other one. While humans are able to disperse resources like this, uniform dispersion rarely occurs in nature. The final type of dispersion is called random. Here, resources are haphazardly placed about. While there are environments where this does occur, they are not common. Figure 10.5 details the three types of dispersion patterns.

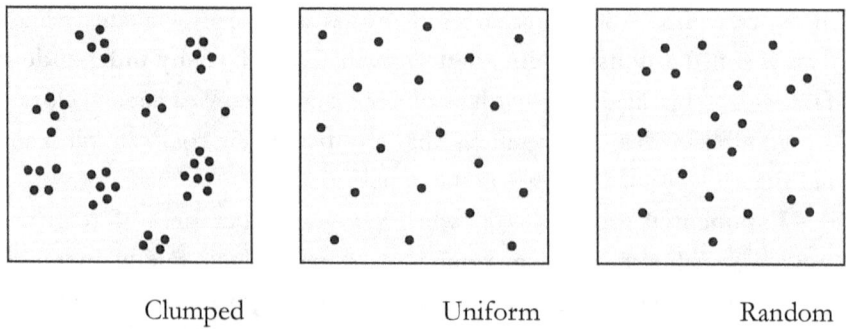

Clumped Uniform Random

Figure 10.5. Dispersion Patterns of Resources

Communities

To survive and reproduce, each organism must have the appropriate behaviors in its repertoire and a habitat that provides its needs. Organisms don't live in isolation; they live in communities of other organisms of different species.

Niches

A **niche** describes the role of a species or a population in an ecosystem. It includes how a population responds to the relative abundance of resources and enemies (e.g., by growing when resources are abundant and predators, parasites, and pathogens are scarce). A niche also indicates the life history of an organism, its habitat, and its place in the food web.

The full range of environmental conditions (biological and physical) under which an organism can exist describes its **fundamental niche**. Because of pressure from superior competitors, organisms often are driven to occupy a niche much narrower than their previous niche. This is known as the **realized niche**.

The following are examples of niches:

Oak trees

- live in forests
- absorb sunlight by photosynthesis
- provide shelter for many animals
- act as support for creeping plants
- serve as a source of food (acorns) for animals
- cover the ground with dead leaves in the autumn

If the oak trees were cut down or destroyed by fire or storms, they would no longer be able to provide shelter, support, and food, which would have a disastrous effect on all the other organisms living in the same habitat.

Hedgehogs

- live near or in gardens
- eat a variety of insects and other invertebrates that live underneath dead leaves and twigs
- have spines that provide a superb environment for fleas and ticks
- put the nitrogen back into the soil when they urinate
- eat slugs, thereby protecting garden plants

If hedgehogs ceased to exist, the slug population would explode and the nutrients in the dead leaves and twigs would not be recycled.

Interspecific relationships

Symbiosis occurs when two species live close together. Sometimes these relationships are beneficial to both organisms, while other times one organism can be harmed. There are four main classifications of these relationships.

Commensalism occurs when one species benefits from the other without causing any harm to the other species. In the ocean, the great whales often migrate from their cooler summer feeding grounds to the warmer winter breeding grounds. They make this same migration every year. Often attached to the whales are barnacles. These small arthropods settle out of the water as juveniles onto the whales and live there for their entire lives. The barnacles, which are usually attached to rocks on the shore, get a free ride from the whales. This increases their chances of finding food. The whales, on the other hand, are completely unphased by the presence of the barnacles. They probably do not even know they are there.

Mutualism occurs when both species benefit from one another. Species involved in mutualistic relationships must co-evolve to survive. As one species evolves, the other must as well if it is to be successful in life. The grouper fish and a species of shrimp live in a mutualistic relationship. The shrimp feed off parasites living on the grouper. Thus, the shrimp are fed and the grouper stays healthy. Many microorganisms exist in mutualistic relationships.

Predation and **parasitism** are beneficial for one species and detrimental for the other. Predation occurs when one organism eats another. The common conception of predation is of a carnivore consuming other animals. This is one form of predation. Although not always resulting in the death of the plant, herbivory is also a form of predation. Some animals eat enough of a plant to cause death. **Parasitism** occurs when a predator lives on or in its host, causing detrimental effects to the host. A tapeworm living within the intestines of large mammals is an example of a parasite. The tapeworm is ingested by the host as an egg in contaminated water. The egg travels through the host's digestive system unharmed until it reaches the small intestine. Here it hatches and grows (sometimes to great lengths). Since the small intestine is the primary site of nutrient absorption, the tapeworm prevents the host from getting the nutrients it needs to survive. The host may be eating constantly, but it would still be losing weight. Over time, the parasite may kill the host. Before it does, however, it will take measures to ensure its survival.

Ecosystems and biomes

An ecosystem includes all living things in a specific area and the non-living things that affect them. The term **biome** is usually used to classify the general types of ecosystems in the world. Each biome contains distinct organisms best adapted to that environment. Biomes are also specific to the different latitudes on the globe. For instance, tropical rainforests do not exist at the poles and tundra cannot exist at the equator. Changes in altitude can also impact the type of biome. There may be snow and ice at the top of a mountain, but a tropical rainforest at the bottom. The major terrestrial biomes are desert, grassland, tundra, boreal forest, tropical rainforest, and temperate forest.

Desert

A desert is any location that receives less than 50 cm of precipitation a year. Despite their lack of water, the soils, though loose and silty, tend to be rich. Specialized plants and animals populate deserts that are adapted to using little water or storing it well. Plant species include succulents, such as cacti that store water in leaves and stems. Animals tend to be non-mammalian and small (e.g., lizards and snakes). Large animals are not able to find sufficient shade in the desert and most mammals are not well adapted to store water and withstand heat. Deserts are dry and may be either cold or hot. Hot and dry deserts are what we typically envision when we think of a desert. Hot deserts are located in northern Africa, southwestern United States, and the Middle East. Similarly, cold deserts have little vegetation, small animals, and are located exclusively in Antarctica, Greenland, much of central Asia, and the Arctic. Both types of deserts do receive precipitation in the winter, though it is in the form of rain or fog in hot deserts and the form of snow in the cold.

Grassland

As the name suggests, grasslands include large expanses of grass with only a few shrubs or trees. There are both tropical and temperate grasslands.

Tropical grasslands cover much of Australia, South America, and India. The weather is warm year-round with moderate rainfall, but the rainfall is concentrated in half the year and drought and fires are common in the other half of the year. These fires serve to renew rather than destroy areas within tropical grasslands. This type of grassland supports a large variety of animals from insects to mammals such as mice, gophers, giraffes, zebras, kangaroos, lions, and elephants.

Temperate grasslands receive less rain than tropical grasslands and are found in South Africa, Eastern Europe, and the western United States. As in tropical grasslands, periods of drought and fire serve to renew the ecosystem. Differences in temperature also differentiate the temperate from the tropical grasslands. Temperate grasslands are cooler in general and experience even colder temperatures in winter. These grasslands support similar types of animals as the tropical grasslands: prairie dogs, deer, mice, coyotes, hawks, snakes, and foxes.

The savanna is grassland with scattered individual trees. Plants of the savanna include shrubs and grasses. Temperatures range from 0°–25° C depending on location. Rainfall is from 90 to 150 cm per year. The savanna is a transitional biome between the rain forest and the desert that is located in central South America, southern Africa, and parts of Australia.

Tundra

The tundra is a treeless plain with extremely low temperatures (-28 to 15° C) and little vegetation or precipitation. Rainfall is limited, ranging from 10 to 15 cm per year. A layer of permanently frozen subsoil, called **permafrost**, is found in the tundra. Permafrost prevents vegetation with deep root systems from surviving, but low shrubs, mosses, grasses, and lichen can live in the shallow soil. These plants grow low and close together to resist the cold temperature and strong winds. The few animals that live in the tundra are adapted to the cold winters (via layers of subcutaneous fat, hibernation, or migration) and raise their young quickly during the short summer. Such species include lemmings, caribou, arctic squirrels and foxes, polar bears, cod, and salmon.

Forests

There are three types of forests, all characterized by the abundant growth of trees, but with different climates, flora, and fauna.

Boreal forest (taiga)

These forests are located throughout northern Europe, Asia, and North America, near the poles. The climate typically consists of short, rainy summers followed by long, cold winters with snow. The trees in boreal forests are adapted to cold winters and are typically evergreens including pine, fir, and spruce. The trees are so thick that there is little undergrowth. A number of animals are adapted to life in the boreal forest, including many mammals

such as bear, moose, wolves, chipmunks, weasels, mink, and deer. These coniferous forests have temperatures ranging from -24 to 22° C. Rainfall is 35–40 cm per year. This is the largest terrestrial biome.

Tropical rainforest

Tropical rainforests are located near the equator and are typically warm and wet throughout the entire year. The temperature is constant (25° C) and the length of daylight is about 12 hours. Precipitation is frequent and occurs evenly throughout the year (hence the name "rainforest"). In a tropical rainforest, rainfall exceeds 200 cm per year. Tropical rainforests have abundant, diverse species of plants and animals. A tropical dry forest gets scarce rainfall, and a tropical deciduous forest has wet and dry seasons. The soil is surprisingly nutrient-poor, and most of the biomass is located within the trees themselves. Vegetation is highly diverse, including many trees with shallow roots, orchids, vines, ferns, mosses and palms. Similarly, animals are plentiful and include all type of birds, reptiles, bats, insects and small to medium-sized mammals.

Temperate forest

These forests have well defined winters and summers with precipitation throughout the year. Temperate forests are common in western Europe, eastern North America, and parts of Asia. Common trees include deciduous species such as oak, beech, maple, and hickory. Unlike the boreal forests, the canopy in temperate forests is not particularly heavy so various smaller plants occupy the understory. Mammals and birds are the predominate form of animal life. Typical species include squirrels, rabbits, skunks, deer, bobcats, and bear. The temperature here ranges from -24 to 38° C. Rainfall is 65–150 cm per year.

A subtype of temperate forests is the **chaparral forest**. Chaparral forests experience mild, rainy winters and hot, dry summers. Trees do not grow as well here. Spiny shrubs dominate. Regions of chaparral forests include the Mediterranean, California coastline, and southwestern Australia.

Aquatic ecosystems

Aquatic ecosystems are, as the name suggests, ecosystems located within bodies of water. Aquatic biomes are divided between fresh water and marine systems. Freshwater ecosystems are closely linked to terrestrial biomes. Lakes, ponds, rivers, streams, and swamplands are examples of freshwater biomes.

Marine ecosystems encompass coastal estuaries, coral reefs (including the largest living structure on Earth—the Great Barrier Reef), and the open sea.

Ponds and Lakes

Varied ecosystems are found in ponds and lakes. Some lakes are even seasonal, lasting only a few months each year. In addition, within lakes there are zones, comparable to those in oceans. The littoral zone, located near the shore and at the top of the lake, is the warmest and lightest zone. Organisms in this zone typically include aquatic plants and insects, snails, clams, fish, and amphibians.

Further from land, but still at the surface of the lake is the limnetic zone. Plankton are abundant in the limnetic zone and are at the bottom of the food chain in this zone, ultimately supporting all freshwater fish. Deeper in the lake is the profundal zone, which is cooler and darker. As plankton die they fall to the bottom of the lake and therefore also serve as a valuable food source for small fish

Rivers and Streams

This biome includes moving bodies of water whose organisms vary according to latitude and geological features. Additionally, characteristics of the stream change as it flows from its headwaters to the sea. Rivers also have depth zones. Different species live in the upper, sunlit areas (e.g., algae, top feeding fish, and aquatic insects) and in the darker, bottom areas (e.g., catfish, carp, and microbes).

Wetlands

Wetlands are the only aquatic biome that is partially land-based. They are areas of standing water in which aquatic plants grow. These species, called hydrophytes, are adapted to extremely humid and moist conditions and include lilies, cattails, sedges, cypress, and black spruce. Animal life in wetlands includes insects, amphibians, reptiles, many birds, and a few small mammals. Though wetlands are usually classified as a freshwater biome, they are in fact salt marshes that support shrimp, various fish, and grasses.

Oceans

Marine areas cover 75% of the Earth's surface and comprise the largest biome on the planet. This biome is organized by depth and temperature of the water. Within the world's oceans, there are several separate zones, each with its own temperature profile and unique species.

The open ocean is very nutrient poor compared to those areas closer to land and to land itself. Temperatures do not vary much because of the specific heat property of water. There is also such a great volume of water that the heat is fairly well distributed. Other than in the far north, it is not common to see the oceans freeze.

While not the most nutrient-rich biome, due to its size, oceans are responsible for most of the world's food and oxygen. Microscopic algae called diatoms inhabit all of the world's oceans and are champion photosynthesizers.

Coral reefs

Coral reefs are located in warm, shallow water near large land masses. The most well-known example is the Great Barrier Reef off the coast of Australia. The coral itself is the dominant life form in the reefs and obtains its nutrients largely through photosynthesis (performed by the algae that live within the coral polyps). Many other animal life forms also populate coral reefs such as fish, octopuses, sea stars, and urchins.

Estuaries

Estuaries are found where fresh and seawater meet, for instance where rivers flow into the oceans. Many species have evolved to thrive in the unique salt concentrations that exist in estuaries. These species include marsh grasses, mangrove trees, oysters, crabs, and certain waterfowl.

Succession

Succession is an orderly process of replacing a community that has been damaged or has previously ceased to exist. Imagine a volcano has just erupted and the flowing lava has wiped out all of the trees, plants, and animals in the area. All that is left is a barren wasteland covered with rock. Over time, the rocks get covered with lichens. These are called the **pioneer species** (the first ones to move into an area after a disruption). The lichens photosynthesize to make energy for themselves, but they also secrete acids that break down the rocks. As the lichen die, their decomposing bodies add nutrients to the rock particles turning it into soil. As time goes on, small grasses and other plants will come into the area and take root. With them come insects. As the soil becomes more fertile, larger and larger plants will come in, outcompeting the smaller ones for light. Ultimately, large trees will grow here. Along with the animals that have moved in, they will form the **climax community**. This process describes

primary succession (see Figure 10.6). The key to primary succession is that no life existed in the area before the pioneer species move in.

Stages of Forest Succession

| Pioneer Plants (Annual) | Perennial Plants and Grasses | Shrubs (Woody Pioneers) | Short-lived Pioneer Trees (Young Forest) | Climax Forest (Mature Forest) |

Time ---→

Figure 10.6. Primary Succession

Secondary succession is very similar to primary succession, but it takes place in communities that were once flourishing but were disturbed by some source, either humans or nature, but not totally stripped. For example, if a farmer stops plowing her fields for several years, many small trees and other plants will grow. Also, many types of insects and other animals will move in.

People are Doing Bad Things: Human Impact on the Planet

The world's population is somewhere in the range of seven and a half billion people and is expected to continue to increase in the near future. All of these people need resources. Most importantly, they need food and places to live. Because of better health, people are living longer, so demand for these resources is never ending.

You Don't Belong Here – Invasive Species

As the ability to move materials all over the world has become easier, so has the possibility that new species can be brought from one place to another. These new organisms are called **invasive species**. They are often better competitors than the native species and have no natural predators in the area. As a result, their populations tend to explode, taking over the areas where they live. They also usually have wider climate tolerances, so they are

able to expand their ranges without much difficulty.

Many times produce imported from tropical regions and ship ballast water are the means for bringing new species into an area. For example, the Asian shore crab was introduced into the New Jersey ecosystem in 1988. It is believed to have arrived in the ballast of a cargo ship delivering materials from Asia. Since then, it has expanded its range up and down the east coast. This crab is an opportunistic omnivore, which means that it will eat just about anything. Its presence has caused the localized extinction of several native species of crabs. Research is currently underway to find a way to control its ever-expanding population.

Resources and their Effects

Humans are continuously searching for new places to live. This encroachment leads to the destruction of natural wildlife communities. They clear-cut trees and fill in lakes to make room to build houses. However, these actions also destroy habitat. Some species are so specialized in in their habitat requirements that any change can cause localized extinction. Conservationists focus on endangered species, but the primary focus should be on protection of the entire biome. If a biome becomes extinct, wildlife will die or invade another biome. Reservations established by governments aim at protecting small parts of biomes. While beneficial in the conservation of a few areas, the majority of the environment is still unprotected.

Humans are responsible for the depletion of the ozone layer that protects the Earth from the majority of incoming UV radiation. This depletion is due to the chemicals used for refrigeration and aerosols. The consequences of ozone depletion include increased UV radiation hitting the Earth, which increases the incidence of skin cancer and has numerous other unknown effects on wildlife and plants.

Humans have had a tremendous impact on the world's natural resources. The world's natural water supplies are affected by human use. Waterways are major sources for recreation and freight transportation. Oil and waste from boats and cargo ships pollute the aquatic environment adversely impacting aquatic animal and plant life, especially coastal fish nurseries and delicate coral reefs. Overfishing worldwide has depleted a major source of protein.

Human's need for fuel has led to substantial depletions of coal and oil reserves and also caused changes in the climate. The constant emissions from

factories, cars, and other places that burn fossil fuels has created a layer of gases that insulates the planet. Called greenhouse gases, these prevent solar radiation from leaving the Earth. Instead, it gets trapped in the atmosphere, causing an overall warming of the planet. Increased surface temperatures result in melting of the polar ice caps, increased ocean levels, habitat loss, and dramatic changes in the weather (can you say El Niño?).

Because of the increasing scarcity of traditional energy sources, scientific research into energy production and management is increasingly focusing on the creation of alternative, efficient means of energy production. Examples of potential sources of alternative energy include wind, water, solar, nuclear, geothermal, and biomass. An important concern in the production and use of energy, from both traditional and alternative sources, is the effect on the environment and the safe disposal of waste products.

CLEP Biology Sample Exam 1

None of the following sample questions has appeared on an actual CLEP examination. They are intended to show you the types of questions and level of difficulty that you will encounter on the actual examination. The exam will indicate topics you may need to study further, and the questions themselves will provide you with content for practice and review. Keep in mind that knowing the answers to all of the sample questions does not a guarantee that you will pass the CLEP Biology exam.

Directions: Each question has five possible answers. For each, choose the one that you feel best answers the question or completes the statement.

1. **Which is not true about a cell membrane?**

 [A] It is made from phospholipids.

 [B] Both plant and animal cells have a cell membrane.

 [C] The cell wall is the same as the cell membrane in plants.

 [D] It controls the passage of nutrients within a cell.

 [E] It contains embedded proteins that help with passage.

2. **Microorganisms use all but which of the following for locomotion?**

 [A] Pseudopods

 [B] Flagella

 [C] Cilia

 [D] Pili

 [E] Villi

3. Which of the following does not possess eukaryotic cells?

 [A] Bacteria

 [B] Protists

 [C] Fungi

 [D] Animals

 [E] Plants

4. Which of the following groups of organisms is composed of those with one cell and no nuclear membrane?

 [A] Monera

 [B] Protista

 [C] Fungi

 [D] Algae

 [E] Plantae

5. Which of the following are found on the outside of the rough endoplasmic reticulum?

 [A] Vacuoles

 [B] Mitochondria

 [C] Microfilaments

 [D] Ribosomes

 [E] Flagella

6. **Identify the correct sequence of organization of living things.**

 [A] Cell – organelle – organ – tissue – organ system – organism

 [B] Cell – tissue – organ – organelle – organ system – organism

 [C] Organelle – cell – tissue – organ – organ system – organism

 [D] Organ system – tissue – organelle – cell – organism – organ

 [E] Organism – organ system – tissue – cell – organelle – organ

7. **Which of the following is not a characteristic shared by all living things?**

 [A] Movement

 [B] Made of cells

 [C] Metabolism

 [D] Reproduction

 [E] Respond to stimuli

8. **What is the purpose of the Golgi apparatus?**

 [A] To break down proteins

 [B] To break down fats

 [C] To make carbohydrates

 [D] To provide the cell with energy

 [E] To sort, modify and package molecules

9. What do amyloplasts do?

 [A] Store starch in a plant cell

 [B] Remove waste in animal cells

 [C] Produce green and yellow pigment

 [D] Aid in photosynthesis.

 [E] Provide energy for metabolism

10. Which of the following does not belong to the domain Archaea?

 [A] Methanogens

 [B] Extreme halophiles

 [C] Thermoacidophiles

 [D] Bacteriophiles

 [E] Sulfobales

11. The first cells that evolved on Earth were probably of which type?

 [A] Autotrophic

 [B] Eukaryotic

 [C] Heterotrophic

 [D] Prokaryotic

 [E] Endosymbiotic

12. During which part of photosynthesis is oxygen given off?

 [A] Light reactions

 [B] Dark reactions

 [C] Krebs cycle

 [D] Reduction of NAD+ to NADH

 [E] Phosphorylation

13. Bacteria commonly reproduce by a process called binary fission. Which of the following best defines this process?

 [A] Viral vectors carry DNA to new bacteria.

 [B] DNA from one bacterium enters another.

 [C] DNA doubles and the bacterial cell divides.

 [D] DNA from dead cells is absorbed into bacteria.

 [E] Bacteria merge with others to form new species.

14. Which tool is best for studying the individual parts of cells?

 [A] Ultracentrifuge

 [B] Phase-contrast microscope

 [C] CAT scan

 [D] Electron microscope

 [E] Light microscope

15. Which of the following classifications includes the thermoacidophiles?

 [A] Plantae

 [B] Animalia

 [C] Bacteria

 [D] Protista

 [E] Archaea

16. Which of the following is not part of the cytoskeleton?

 [A] Vacuoles

 [B] Microfilaments

 [C] Microtubules

 [D] Intermediate filaments

 [E] Motor proteins

17. Of what are viruses made?

 [A] A protein coat surrounding a nucleic acid

 [B] RNA and protein surrounded by a cell wall

 [C] A nucleic acid surrounding a protein coat

 [D] Protein surrounded by DNA

 [E] A lipid bilayer surrounding a protein coat and RNA

18. Which of the following are used to classify protists into their major groups?

 [A] Their method of obtaining nutrition

 [B] Their method of reproduction

 [C] Their use of metabolism

 [D] Their form and function

 [E] Their means of locomotion

19. Replication of chromosomes occurs during which phase of the cell cycle?

 [A] Prophase

 [B] Interphase

 [C] Metaphase

 [D] Anaphase

 [E] Telophase

20. Which of the following events occurs during telophase in a plant cell?

 [A] The chromosomes are doubled.

 [B] A cell plate forms.

 [C] Crossing over occurs.

 [D] A cleavage furrow develops.

 [E] Spindle fibers become visible.

21. Which stage of mitosis is seen in the diagram?

[A] Anaphase

[B] Metaphase

[C] Telophase

[D] Prophase

[E] Interphase

22. Which stage of mitosis is shown in the diagram?

[A] Prophase

[B] Telophase

[C] Anaphase

[D] Metaphase

[E] Interphase

23. Which stage of mitosis is shown in the diagram?

[A] Interphase

[B] Metaphase

[C] Prophase

[D] Telophase

[E] Anaphase

24. Which of the following is a monomer?

[A] RNA

[B] Glycogen

[C] DNA

[D] Amino acid

[E] Lipid

25. Which of the following does not affect the reaction rate of an enzyme?

[A] Increase of temperature

[B] Amount of substrate

[C] pH

[D] Size of the cell

[E] Concentration of enzyme

26. **All but which one of the following is true of a cell membrane?**

 [A] It contains polar and nonpolar phospholipids.

 [B] It only uses active transport to move molecules across it.

 [C] It contains cholesterol.

 [D] It has proteins imbedded within it.

 [E] It is selectively permeable to many substances.

27. **Which of the following describes facilitated diffusion?**

 [A] It requires energy.

 [B] It only happens in plant cells.

 [C] It only allows molecules to leave a cell but not to enter it.

 [D] It produces a significant amount of energy for the cell.

 [E] It needs a transport molecule to pass through the membrane.

28. **Which of the following is not true of enzymes?**

 [A] They are the most diverse of all proteins.

 [B] They act on a substrate.

 [C] They work at a wide range of pH.

 [D] They are temperature-dependent.

 [E] They have specialized functions.

29. Which of the following is necessary for diffusion to occur?

 [A] Carrier proteins

 [B] Energy

 [C] Water molecules

 [D] A cell membrane

 [E] A concentration gradient

30. Which of the following is an example of the use of energy to move a substance through a membrane from areas of low concentration to areas of high concentration?

 [A] Osmosis

 [B] Active transport

 [C] Exocytosis

 [D] Phagocytosis

 [E] Facilitated diffusion

31. A plant cell is placed in salt water. What is the resulting movement of water out of the cell called?

 [A] Facilitated diffusion

 [B] Diffusion

 [C] Transpiration

 [D] Osmosis

 [E] Active transport

32. What are the monomers of polysaccharides?

 [A] Nucleotides

 [B] Amino acids

 [C] Polypeptides

 [D] Fatty acids

 [E] Simple sugars

33. Which type of cell would contain the most mitochondria?

 [A] Muscle cell

 [B] Nerve cell

 [C] Epithelial cell

 [D] Blood cell

 [E] Bone cell

34. According to the fluid-mosaic model of the cell membrane, what are membranes composed of?

 [A] Phospholipid bilayers with proteins embedded in the layers

 [B] One layer of phospholipids with cholesterol embedded in the layer

 [C] Two layers of protein with lipids embedded in the layers

 [D] DNA and fluid proteins with carbohydrates embedded in the layer

 [E] Glycerol and RNA with carbohydrates embedded in the layer

35. **Which is the correct statement regarding the human nervous system and the human endocrine system?**

 [A] The nervous system maintains homeostasis whereas the endocrine system does not.

 [B] Endocrine glands produce neurotransmitters whereas nerves produce hormones.

 [C] Nerve signals travel on neurons whereas hormones travel through the blood.

 [D] The nervous system involves chemical transmission whereas the endocrine system does not.

 [E] The nervous system produces physiological responses whereas the endocrine produces behavioral.

36. **Which process generates the most ATP?**

 [A] Fermentation

 [B] Glycolysis

 [C] The Calvin cycle

 [D] The Krebs cycle

 [E] Chemiosmosis

37. **Which of the following is a function of the cardiovascular system?**

 [A] Move oxygenated blood around the body

 [B] Oxygenate the blood through gas exchange

 [C] Act as an exocrine system

 [D] Flush toxins out of the body

 [E] Transport signals from the brain

38. Which of the following is not a part of the nervous system?

 [A] Brain

 [B] Spinal cord

 [C] Axons

 [D] Venules

 [E] Cochlea

39. Organisms need to maintain a constant internal environment to survive. Which of the following is a process by which they achieve this?

 [A] Respiration

 [B] Reproduction

 [C] Depolarization

 [D] Repolarization

 [E] Thermoregulation

40. Which of the following controls the body's endocrine mechanisms?

 [A] Feedback loops

 [B] Control molecules

 [C] Neurochemicals

 [D] Neurotransmitters

 [E] Behavioral responses

41. Which gland regulates the calcium in the body?

 [A] Thyroid gland

 [B] Parathyroid gland

 [C] Hypothalamus

 [D] Pituitary gland

 [E] Pancreas

42. Which of the following steroids is not created in the gonads?

 [A] Testosterone

 [B] Estrogen

 [C] Progesterone

 [D] ACTH

 [E] FSH

43. What is the most common neurotransmitter?

 [A] Epinephrine

 [B] Serotonin

 [C] Acetylcholine

 [D] Norepinephrine

 [E] Oxytocin

44. Food is carried through the digestive tract by a series of wave-like contractions. What is this process is called?

 [A] Peristalsis

 [B] Chyme

 [C] Digestion

 [D] Absorption

 [E] Depolarization

45. Which of the following must muscles pull on in order to initiate movement?

 [A] Skin

 [B] Bones

 [C] Joints

 [D] Ligaments

 [E] Bursa

46. Hormones are essential to the regulation of reproduction. Which organ is responsible for the release of hormones for sexual maturity?

 [A] Pituitary gland

 [B] Hypothalamus

 [C] Pancreas

 [D] Thyroid gland

 [E] Pineal gland

47. What is the type of muscle in the human body that is voluntary?

 [A] Cardiac

 [B] Sarcomere

 [C] Smooth

 [D] Skeletal

 [E] Actin

48. The wrist is an example of what kind of joint?

 [A] Ball and socket

 [B] Pivot

 [C] Stationary

 [D] Hinge

 [E] Gliding

49. What is the waterproofing protein in the skin called?

 [A] Actin

 [B] Epidermis

 [C] Collagen

 [D] Sebum

 [E] Keratin

50. What is small flap of tissue called that covers the trachea when swallowing food?

 [A] Epiglottis

 [B] Larynx

 [C] Connective tissue

 [D] Villi

 [E] Squamous tissue

51. What is the role of neurotransmitters in nerve action?

 [A] To turn off the sodium pump

 [B] To turn off the calcium pump

 [C] To send impulses to neurons

 [D] To send impulses around the body

 [E] To send impulses from axon to dendrite

52. Which substance breaks down fats?

 [A] Bile produced in the gall bladder

 [B] Lipase produced in the gall bladder

 [C] Glucagons produced in the liver

 [D] Amylase produces in the gall bladder

 [E] Bile produced in the liver

53. Where does fertilization in humans usually occur?

 [A] Uterus

 [B] Ovary

 [C] Fallopian tubes

 [D] Vagina

 [E] Epididymis

54. Which of the following is lacking in the dermis layer of skin?

 [A] Sweat glands

 [B] Keratin

 [C] Hair follicles

 [D] Blood vessels

 [E] Living cells

55. A school age boy had the mumps as a baby. Why will he most likely not get this disease again?

 [A] Passive immunity

 [B] Vaccination

 [C] Antibiotics

 [D] Active immunity

 [E] Antigen production

56. **What is any foreign particle called that causes an immune reaction?**

 [A] An antigen

 [B] A histocompatibity complex

 [C] An antibody

 [D] A vaccine

 [E] A bacteriophage

57. **Which of the following statements describes the polymerase chain reaction?**

 [A] It is a group of polymerases.

 [B] It is a technique for amplifying DNA.

 [C] It is a primer for DNA synthesis.

 [D] It is a way to synthesize polymerase.

 [E] It is a series of genetic mutations.

58. **Which part of a DNA nucleotide can vary?**

 [A] Deoxyribose

 [B] Phosphate group

 [C] Hydrogen bonds

 [D] Sugar

 [E] Nitrogenous base

59. A DNA strand has the base sequence of TCAGTA. Its DNA complement would have which of the following sequences?

 [A] ATGACT

 [B] TCAGTA

 [C] AGUCAU

 [D] AGTCAT

 [E] TCTGTA

60. Which of the following carries amino acids to the ribosome during protein synthesis?

 [A] Messenger RNA

 [B] Ribosomal RNA

 [C] Transfer RNA

 [D] DNA

 [E] RNA

61. A protein is 60 amino acids in length. This requires a coded DNA sequence of how many nucleotides?

 [A] 20

 [B] 30

 [C] 120

 [D] 180

 [E] 240

62. A DNA molecule has the sequence of ACTATG. What is the anticodon of this molecule?

 [A] UGAUAC

 [B] ACUAUG

 [C] TGATAC

 [D] ACTATG

 [E] CTGCGA

63. What is the general term for a change that affects the sequence of bases in a gene?

 [A] Deletion

 [B] Polyploid

 [C] Mutation

 [D] Duplication

 [E] Substitution

64. Segments of DNA can be transferred from the DNA of one organism to another through the use of which of the following?

 [A] Bacterial plasmids

 [B] Viruses

 [C] Chromosomes from frogs

 [D] Plant DNA

 [E] Okazaki fragments

65. What is the enzyme that unwinds DNA during replication?

 [A] DNAse

 [B] DNA replicase

 [C] DNA helicase

 [D] DNA topoisomerases

 [E] DNA polymerase

66. What is a small circular piece of DNA called that contains accessory DNA?

 [A] Mitochondrial DNA

 [B] Messenger RNA

 [C] Transfer DNA

 [D] Okazaki fragment

 [E] Plasmid

67. In DNA, adenine bonds with _____, while cytosine bonds with _____.

 [A] Thymine/guanine

 [B] Adenine/cytosine

 [C] Cytosine/uracil

 [D] Guanine/thymine

 [E] Uracil/adenine

68. Which protein structure consists of the coils and folds of polypeptide chains?

 [A] Secondary structure

 [B] Quaternary structure

 [C] Tertiary structure

 [D] Primary structure

 [E] Quinary structure

69. What can be said about homozygous individuals?

 [A] They have two different alleles.

 [B] They are of the same species.

 [C] They exhibit the same features.

 [D] They have a pair of identical alleles.

 [E] They produce identical offspring.

70. The term "phenotype" refers to which of the following?

 [A] A condition that is heterozygous

 [B] The genetic makeup of an individual

 [C] A condition that is homozygous

 [D] The expression of the genotype

 [E] From which parent the traits were inherited

71. The ratio of brown-eyed to blue-eyed children from the mating of a blue-eyed male to a heterozygous brown-eyed female is expected to be which of the following?

 [A] 3:1

 [B] 2:2

 [C] 1:0

 [D] 1:2

 [E] 0:4

72. Which of the following defines the Law of Segregation defined by Gregor Mendel?

 [A] After meiosis, each new cell will contain an allele that is recessive.

 [B] Only one of two alleles is expressed in a heterozygous organism.

 [C] The allele expressed is always the dominant allele.

 [D] Alleles of one trait do not affect the inheritance of alleles on another chromosome.

 [E] When sex cells form, the two alleles that determine a trait will end up on different gametes.

73. Which of the following is an example of the incomplete dominance that occurs when a white flower is crossed with a red flower?

 [A] Pink flowers

 [B] Red flowers

 [C] White flowers

 [D] Red and white flowers

 [E] White and pink flowers

74. A child with type O blood has a father with type A blood and a mother with type B blood. The genotypes of the parents respectively would be which of the following?

 [A] AA and BO

 [B] AO and BO

 [C] AA and BB

 [D] AO and OO

 [E] OO and AB

75. Crossing over, which increases genetic diversity, occurs during which stage(s) of meiosis?

 [A] Telophase II in meiosis

 [B] Metaphase in mitosis

 [C] Interphase in meiosis

 [D] Prophase I in meiosis

 [E] Metaphase II in meiosis

76. ABO blood grouping is an example of which type of allele dominance?

 [A] Autosomal dominance

 [B] Incomplete dominance

 [C] Somatic dominance

 [D] Complete dominance

 [E] Codominance

77. In a Punnett square with a single trait, what are the ratios of genotypes produced between two heterozygous individuals?

 [A] 1:2:2

 [B] 2:1:1

 [C] 1:1:1

 [D] 1:2:1

 [E] 2:2:2

78. What is the term for an organism's genetic makeup?

 [A] Heterozygote

 [B] Genotype

 [C] Phenotype

 [D] Homozygote

 [E] Dominance

79. Which of the following represents a genetic engineering advancement in the medical field?

 [A] Stem cell reproduction

 [B] Pesticides

 [C] Degradation of harmful chemicals

 [D] Antibiotics

 [E] Gene therapy

80. Which of the following is not true regarding restriction enzymes?

 [A] They aid in transcombination procedures.

 [B] They are used in genetic engineering.

 [C] They are named after the bacteria in which they naturally occur.

 [D] They identify and splice certain base sequences on DNA.

 [E] They can be produced by certain lipids during DNA replication.

81. Which of the following processes is not one of the modern uses of DNA?

 [A] PCR technology

 [B] Gene therapy

 [C] Cloning

 [D] Genetic Alignment

 [E] Transgenic organisms

82. Which statement best represents gel electrophoresis?

 [A] It isolates fragments of DNA for scientific purposes.

 [B] It cannot be used in proteins.

 [C] It requires the polymerase chain reaction.

 [D] It only separates DNA by size.

 [E] It uses different charged particles to color the bands.

83. What is the term that describes the duplication of genetic material in another cell?

 [A] Replicating

 [B] Cell duplication

 [C] Transgenics

 [D] Genetic restructuring

 [E] Cloning

84. What does gel electrophoresis use to separate the DNA?

 [A] The amount of current

 [B] The size of the molecule

 [C] The positive charge of the molecule

 [D] The solubility of the gel

 [E] The source of the DNA

85. Which of the following is a result of reproductive isolation?

 [A] Extinction

 [B] Migration

 [C] Fossilization

 [D] Speciation

 [E] Radiation

86. **Which of the following is true about natural selection?**

 [A] It acts on an individual genotype.

 [B] It is not currently happening.

 [C] It is only an animal phenomenon.

 [D] It acts on the individual phenotype.

 [E] It is used to prevent overpopulation.

87. **How does diversity aid a population?**

 [A] Individuals are better able to survive.

 [B] Mates are attracted to a diverse population.

 [C] Potential mates like conformity.

 [D] It increases the DNA differences in the population.

 [E] It provides possible improvements to the population.

88. **Which statement is not true about diversity?**

 [A] Without diversity there would be extinction.

 [B] Diversity is increasing all the time.

 [C] Fossil evidence supports diversity.

 [D] Sexual reproduction encourages more diversity.

 [E] Skeletons are too similar to allow for diversity.

89. Which of the following ideas was a major part of Darwin's evolutionary theory?

 [A] Punctualism

 [B] Gradualism

 [C] Equilibrium

 [D] Convergency

 [E] Altruism

90. Which statement is not true about reproductive isolation?

 [A] It prevents populations from exchanging genes.

 [B] It can occur by preventing fertilization.

 [C] It can result in speciation.

 [D] It happens more often on the mainland.

 [E] It produces offspring with unique phenotypes.

91. Which idea is true about members of the same species?

 [A] They look identical.

 [B] They never change.

 [C] They reproduce successfully within their group.

 [D] They live in the same geographic location.

 [E] They have very dissimilar genotypes.

92. Which of the following factors will affect the Hardy-Weinberg law of equilibrium, leading to evolutionary change?

 [A] No mutations

 [B] Non-random mating

 [C] No immigration or emigration

 [D] Large population

 [E] Small individual species

93. If a population is in Hardy-Weinberg equilibrium and the frequency of the recessive allele is 0.3, what percentage of the population is expected to be heterozygous?

 [A] 9%

 [B] 49%

 [C] 42%

 [D] 21%

 [E] 7%

94. Which aspect of science does not support evolution?

 [A] Comparative anatomy

 [B] Organic chemistry

 [C] Comparison of DNA among organisms

 [D] Analogous structures

 [E] Embryology

95. In which of the following does evolution occur?

 [A] Individuals

 [B] Populations

 [C] Organ systems

 [D] Cells

 [E] Ecosystems

96. Which process contributes most to the large variety of living things in the world today?

 [A] Meiosis

 [B] Asexual reproduction

 [C] Mitosis

 [D] Alternation of generations

 [E] Reproductive isolation

97. Which of the following gases was a major part of the primitive Earth's atmosphere?

 [A] Fluorine

 [B] Methane

 [C] Oxygen

 [D] Krypton

 [E] Argon

98. **What is a major principle of the Endosymbiotic Theory?**

 [A] Birds and dinosaurs share a common ancestor.

 [B] Animals evolved in close relationships with one another.

 [C] Prokaryotes arose from eukaryotes.

 [D] Inorganic compounds are the basis of living things.

 [E] Eukaryotes arose from very simple prokaryotes.

99. **The wing of a bird, the human arm, and the pectoral fin of a whale all have the same bone structure. What are these structures called?**

 [A] Polymorphic structures

 [B] Homologous structures

 [C] Vestigial structures

 [D] Analogous structures

 [E] Allopatric structures

100. **Which of the following is not an abiotic factor?**

 [A] Temperature

 [B] Rainfall

 [C] Soil quality

 [D] Predation

 [E] Wind speed

101. Which of the following is not true about cladistics?

 [A] It is the study of phylogenetic relationships of organisms.

 [B] It involves a branching diagram that uses the development of novel traits to separate groups of organisms.

 [C] It distinguishes between the relative importance of the traits.

 [D] It shows when traits developed with respect to other traits.

 [E] It indicates which organisms are most closely related to each other and what their common ancestors were.

102. If DDT were present in an ecosystem, which of the following organisms would have the highest concentration in its body?

 [A] Herring

 [B] Diatom

 [C] Zooplankton

 [D] Salmon

 [E] Osprey

103. What eats secondary consumers?

 [A] Producers

 [B] Tertiary consumers

 [C] Primary consumers

 [D] Decomposers

 [E] Detritivores

104. Which statement is true about the water cycle?

[A] Two percent of the water is fixed and unavailable.

[B] 75% of available water is groundwater.

[C] The water cycle is driven by the ocean currents.

[D] Surface water percolates up from underground springs.

[E] New water is being added into the cycle all the time.

105. Which statement about the carbon cycle is false?

[A] Ten percent of all available carbon is in the air.

[B] Carbon dioxide is fixed by glycosylation.

[C] Plants fix carbon in the form of glucose.

[D] Animals release carbon through respiration.

[E] Most atmospheric carbon comes from the decay of dead organisms.

106. What is the impact of sulfur oxides and nitrogen oxides in the environment when they react with water?

[A] Ammonia

[B] Acidic precipitation

[C] Sulfuric acid

[D] Global warming

[E] Greenhouse effect

107. Which term is not associated with the water cycle?

 [A] Precipitation

 [B] Transpiration

 [C] Fixation

 [D] Evaporation

 [E] Runoff

108. Which of the following is a density dependent factor that affects a population?

 [A] Temperature

 [B] Rainfall

 [C] Predation

 [D] Soil nutrients

 [E] Wind speed

109. High humidity and temperature stability are present in which of the following biomes?

 [A] Taiga

 [B] Deciduous forest

 [C] Desert

 [D] Tropical rain forest

 [E] Coniferous forest

110. Which trophic level is the most ecologically efficient?

 [A] Decomposers

 [B] Producers

 [C] Tertiary consumers

 [D] Secondary consumers

 [E] Primary consumers

111. From where does the oxygen created in photosynthesis come?

 [A] Carbon dioxide

 [B] Chlorophyll

 [C] Glucose

 [D] Carbon monoxide

 [E] Water

112. Which of the following is true of decomposers?

 [A] Decomposers recycle the carbon accumulated in durable organic material.

 [B] Decomposers take nitrogen out of the soil to use for food.

 [C] Decomposers absorb nutrients from the air to maintain their metabolisms.

 [D] Decomposers belong to the Genus Escherichia.

 [E] Decomposers are able to use the sun to produce their own energy.

113. A clownfish is protected by a sea anemone's tentacles, and in turn, the anemone receives uneaten food from the clownfish. What type of relationship is exemplified by this example?

 [A] Mutualism

 [B] Parasitism

 [C] Commensalism

 [D] Competition

 [E] Amensalism

114. Which of the following is most likely to happen in order for primary succession to occur?

 [A] Nutrient enrichment

 [B] A forest fire

 [C] Bare rock is exposed after a water table recedes

 [D] A housing development is built

 [E] A farmer stops cultivating her fields

115. What is the Mendelian law called that states that only one of the two possible alleles from each parent is passed on to the offspring?

 [A] The Mendelian Law

 [B] The Law of Independent Assortment

 [C] The Law of Segregation

 [D] The Allele Law

 [E] The Law of Dominance and Recessiveness

ANSWER KEY

Question Number	Correct Answer	Your Answer	Question Number	Correct Answer	Your Answer	Question Number	Correct Answer	Your Answer
1.	C		40.	A		79.	E	
2.	E		41.	B		80.	A	
3.	A		42.	D		81.	D	
4.	A		43.	C		82.	A	
5.	D		44.	A		83.	E	
6.	C		45.	B		84.	B	
7.	A		46.	B		85.	D	
8.	E		47.	D		86.	D	
9.	A		48.	B		87.	E	
10.	D		49.	E		88.	E	
11.	D		50.	A		89.	B	
12.	A		51.	C		90.	D	
13.	C		52.	E		91.	C	
14.	D		53.	C		92.	B	
15.	D		54.	B		93.	C	
16.	A		55.	D		94.	B	
17.	A		56.	A		95.	B	
18.	D		57.	B		96.	A	
19.	B		58.	E		97.	B	
20.	B		59.	D		98.	E	
21.	B		60.	C		99.	B	
22.	B		61.	D		100.	D	
23.	E		62.	D		101.	C	
24.	D		63.	C		102.	E	
25.	D		64.	A		103.	B	
26.	B		65.	C		104.	A	
27.	E		66.	E		105.	B	
28.	C		67.	A		106.	B	
29.	E		68.	A		107.	C	
30.	B		69.	D		108.	C	
31.	D		70.	D		109.	D	
32.	E		71.	B		110.	B	
33.	A		72.	E		111.	E	
34.	A		73.	A		112.	A	
35.	C		74.	B		113.	A	
36.	E		75.	D		114.	C	
37.	A		76.	C		115.	B	
38.	D		77.	D				
39.	E		78.	B				

CLEP Biology Sample Exam 1 Explanations

1. **Which is not true about a cell membrane?**

 [A] It is made from phospholipids.

 [B] Both plant and animal cells have a cell membrane.

 [C] The cell wall is the same as the cell membrane in plants.

 [D] It controls the passage of nutrients within a cell.

 [E] It contains embedded proteins that help with passage.

 The answer is C.

 Both plants and animals have cell membranes but plant cells also have an outer cell wall to give it structure.

2. **Microorganisms use all but which of the following for locomotion?**

 [A] Pseudopods

 [B] Flagella

 [C] Cilia

 [D] Pili

 [E] Villi

 The answer is E.

 Microorganisms use pseudopods, pili, flagella and cilia for movement. Villi are structures in the small intestine that increase the surface area for absorption.

3. Which of the following does not possess eukaryotic cells?

 [A] Bacteria

 [B] Protists

 [C] Fungi

 [D] Animals

 [E] Plants

 The answer is A.

 Eukaryotic cells are found in protists, fungi, plants and animals but not in bacteria.

4. Which of the following groups of organisms is composed of those with one cell and no nuclear membrane?

 [A] Monera

 [B] Protista

 [C] Fungi

 [D] Algae

 [E] Plantae

 The answer is A.

 Bacteria are unicellular organisms with no nucleus. Algae are unicellular protists; Fungi and Plantae are multicellular with nucleated cells.

5. **Which of the following are found on the outside of the rough endoplasmic reticulum?**

 [A] Vacuoles

 [B] Mitochondria

 [C] Microfilaments

 [D] Ribosomes

 [E] Flagella

 The answer is D.

 Rough endoplasmic reticulum is defined as such because of the occurrence of ribosomes on its surface.

6. **Identify the correct sequence of organization of living things.**

 [A] Cell – organelle – organ – tissue – organ system – organism

 [B] Cell – tissue – organ – organelle – organ system – organism

 [C] Organelle – cell – tissue – organ – organ system – organism

 [D] Organ system – tissue – organelle – cell – organism – organ

 [E] Organism – organ system – tissue – cell – organelle – organ

 The answer is C.

 An organism, such as a human, is composed of several organ systems such as the circulatory and nervous systems. These organ systems consist of many organs including the heart and the brain. These organs are made of tissue such as cardiac muscle. Tissues are made up of cells, which contain organelles like the mitochondria and the Golgi apparatus.

7. Which of the following is not a characteristic shared by all living things?

 [A] Movement

 [B] Made of cells

 [C] Metabolism

 [D] Reproduction

 [E] Respond to stimuli

 The answer is A.

 Movement is not a characteristic of life. Viruses are considered non-living organisms but have the ability to move from cell to cell in its host organism. A lichen attached to a rock is very much alive but unable to move.

8. What is the purpose of the Golgi apparatus?

 [A] To break down proteins

 [B] To break down fats

 [C] To make carbohydrates

 [D] To provide the cell with energy

 [E] To sort, modify and package molecules

 The answer is E.

 The Golgi apparatus takes molecules from the endoplasmic reticulum and sorts, modifies and packages the molecules for later use by the cell.

9. What do amyloplasts do?

 [A] Store starch in a plant cell

 [B] Remove waste in animal cells

 [C] Produce green and yellow pigment

 [D] Aid in photosynthesis.

 [E] Provide energy for metabolism

 The answer is A.

 Amyloplasts store starch in plant cells

10. Which of the following does not belong to the domain Archaea?

 [A] Methanogens

 [B] Extreme halophiles

 [C] Thermoacidophiles

 [D] Bacteriophiles

 [E] Sulfobales

 The answer is D.

 The Archaea group includes all of the above except Bacteriophiles.

11. **The first cells that evolved on Earth were probably of which type?**

 [A] Autotrophic

 [B] Eukaryotic

 [C] Heterotrophic

 [D] Prokaryotic

 [E] Endosymbiotic

 The answer is D.

 Prokaryotes date back to 3.5 billion years ago in the earliest fossil record. Their ability to adapt to the environment allows them to thrive in a wide variety of habitats.

12. **During which part of photosynthesis is oxygen given off?**

 [A] Light reactions

 [B] Dark reactions

 [C] Krebs cycle

 [D] Reduction of NAD+ to NADH

 [E] Phosphorylation

 The answer is A.

 The conversion of solar energy to chemical energy occurs in the light reactions. Electrons are transferred by the absorption of light by chlorophyll and cause water to split, releasing oxygen as a waste product.

13. **Bacteria commonly reproduce by a process called binary fission. Which of the following best defines this process?**

 [A] Viral vectors carry DNA to new bacteria.

 [B] DNA from one bacterium enters another.

 [C] DNA doubles and the bacterial cell divides.

 [D] DNA from dead cells is absorbed into bacteria.

 [E] Bacteria merge with others to form new species.

 The answer is C.

 Binary fission is the asexual process in which the bacteria divide in half after the DNA doubles. This results in an exact clone of the parent cell.

14. **Which tool is best for studying the individual parts of cells?**

 [A] Ultracentrifuge

 [B] Phase-contrast microscope

 [C] CAT scan

 [D] Electron microscope

 [E] Light microscope

 The answer is D.

 The scanning electron microscope uses a beam of electrons to pass through the specimen. The resolution is about 1000 times greater than that of a light microscope. This allows the scientist to view extremely small objects, such as the individual parts of a cell.

15. Which of the following classifications includes the thermoacidophiles?

 [A] Plantae

 [B] Animalia

 [C] Bacteria

 [D] Protista

 [E] Archaea

 The answer is D.

 Thermoacidophiles, methanogens, and halobacteria are members of the Archaea group.

16. Which of the following is not part of the cytoskeleton?

 [A] Vacuoles

 [B] Microfilaments

 [C] Microtubules

 [D] Intermediate filaments

 [E] Motor proteins

 The answer is A.

 Vacuoles are mostly found in plants and hold stored food and pigments. The other three choices are fibers that make up the cytoskeleton found in both plant and animal cells.

17. **Of what are viruses made?**

 [A] A protein coat surrounding a nucleic acid

 [B] RNA and protein surrounded by a cell wall

 [C] A nucleic acid surrounding a protein coat

 [D] Protein surrounded by DNA

 [E] A lipid bilayer surrounding a protein coat and RNA

 The answer is A.

 Viruses are composed of a protein coat surrounding a nucleic acid, either RNA or DNA.

18. **Which of the following are used to classify protists into their major groups?**

 [A] Their method of obtaining nutrition

 [B] Their method of reproduction

 [C] Their use of metabolism

 [D] Their form and function

 [E] Their means of locomotion

 The answer is D.

 The chaotic status of names and concepts of the higher classification of the protists reflects their great diversity in form, function, and life styles. The protists are often grouped as algae (plant-like), protozoa (animal-like), or fungus-like based on the similarity of their lifestyle and characteristics to these more defined groups.

19. **Replication of chromosomes occurs during which phase of the cell cycle?**

 [A] Prophase

 [B] Interphase

 [C] Metaphase

 [D] Anaphase

 [E] Telophase

 The answer is B.

 Interphase is the stage where the cell grows and copies the chromosomes in preparation for the mitotic phase.

20. **Which of the following events occurs during telophase in a plant cell?**

 [A] The chromosomes are doubled.

 [B] A cell plate forms.

 [C] Crossing over occurs.

 [D] A cleavage furrow develops.

 [E] Spindle fibers become visible.

 The answer is B.

 During plant cell telophase, a cell plate is observed whereas a cleavage furrow is formed in animal cells.

21. Which stage of mitosis is seen in the diagram?

[A] Anaphase

[B] Metaphase

[C] Telophase

[D] Prophase

[E] Interphase

The answer is B.

During metaphase, the centromeres are at opposite ends of the cell, and the chromosomes align with one another in the middle of the cell.

22. Which stage of mitosis is shown in the diagram?

[A] Prophase

[B] Telophase

[C] Anaphase

[D] Metaphase

[E] Interphase

The answer is B.

Telophase is the last stage of mitosis. Here, two nuclei become visible and the nuclear membrane reassembles.

23. Which stage of mitosis is shown in the diagram?

[A] Interphase

[B] Metaphase

[C] Prophase

[D] Telophase

[E] Anaphase

The answer is E.

During anaphase, the centromeres split in half and homologous chromosomes separate.

24. Which of the following is a monomer?

[A] RNA

[B] Glycogen

[C] DNA

[D] Amino acid

[E] Lipid

The answer is D.

A monomer is the simplest unit of structure for a particular macromolecule. Amino acids are the basic units that comprise a protein. RNA and DNA are polymers consisting of nucleotides and glycogen is a polymer consisting of many molecules of glucose.

25. **Which of the following does not affect the reaction rate of an enzyme?**

 [A] Increase of temperature

 [B] Amount of substrate

 [C] pH

 [D] Size of the cell

 [E] Concentration of enzyme

 The answer is D.

 Temperature and pH can affect the rate of reaction of an enzyme. The amount of substrate affects the enzyme as well. The enzyme acts on the substrate. The more substrate, the slower the enzyme rate. Therefore, the only choice left is D, the size of the cell, which has no effect on enzyme rate.

26. **All but which one of the following is true of a cell membrane?**

 [A] It contains polar and nonpolar phospholipids.

 [B] It only uses active transport to move molecules across it.

 [C] It contains cholesterol.

 [D] It has proteins imbedded within it.

 [E] It is selectively permeable to many substances.

 The answer is B.

 Cell membranes use passive and active transport to transport molecules across the membrane.

27. Which of the following describes facilitated diffusion?

 [A] It requires energy.

 [B] It only happens in plant cells.

 [C] It only allows molecules to leave a cell but not to enter it.

 [D] It produces a significant amount of energy for the cell.

 [E] It needs a transport molecule to pass through the membrane.

 The answer is E.

 Facilitated diffusion requires no energy but needs a transport molecule to pass another molecule through the membrane.

28. Which of the following is not true of enzymes?

 [A] They are the most diverse of all proteins.

 [B] They act on a substrate.

 [C] They work at a wide range of pH.

 [D] They are temperature-dependent.

 [E] They have specialized functions.

 The answer is C.

 Enzymes generally work best within a very narrow range in pH.

29. **Which of the following is necessary for diffusion to occur?**

 [A] Carrier proteins

 [B] Energy

 [C] Water molecules

 [D] A cell membrane

 [E] A concentration gradient

 The answer is E.

 Diffusion is the ability of molecules to move from areas of high concentration to areas of low concentration (a concentration gradient).

30. **Which of the following is an example of the use of energy to move a substance through a membrane from areas of low concentration to areas of high concentration?**

 [A] Osmosis

 [B] Active transport

 [C] Exocytosis

 [D] Phagocytosis

 [E] Facilitated diffusion

 The answer is B.

 Active transport can move substances with or against the concentration gradient. This energy-requiring process allows for molecules to move from areas of low concentration to areas of high concentration.

31. **A plant cell is placed in salt water. What is the resulting movement of water out of the cell called?**

 [A] Facilitated diffusion

 [B] Diffusion

 [C] Transpiration

 [D] Osmosis

 [E] Active transport

 The answer is D.

 Osmosis is simply the diffusion of water across a semi-permeable membrane. Water will diffuse out of the cell if there is a lower concentration of water on the outside of the cell.

32. **What are the monomers of polysaccharides?**

 [A] Nucleotides

 [B] Amino acids

 [C] Polypeptides

 [D] Fatty acids

 [E] Simple sugars

 The answer is E.

 The monomers of polysaccharides are simple sugars.

33. **Which type of cell would contain the most mitochondria?**

 [A] Muscle cell

 [B] Nerve cell

 [C] Epithelial cell

 [D] Blood cell

 [E] Bone cell

 The answer is A.

 Mitochondria are the site of cellular respiration where ATP is made. Muscle cells have the most mitochondria because they use a great deal of energy.

34. **According to the fluid-mosaic model of the cell membrane, what are membranes composed of?**

 [A] Phospholipid bilayers with proteins embedded in the layers

 [B] One layer of phospholipids with cholesterol embedded in the layer

 [C] Two layers of protein with lipids embedded in the layers

 [D] DNA and fluid proteins with carbohydrates embedded in the layer

 [E] Glycerol and RNA with carbohydrates embedded in the layer

 The answer is A.

 Cell membranes are composed of two phospholipids with their hydrophobic tails sandwiched between their hydrophilic heads, creating a lipid bilayer. The membrane contains proteins embedded in the layer (integral proteins) and proteins on the surface (peripheral proteins).

35. **Which is the correct statement regarding the human nervous system and the human endocrine system?**

 [A] The nervous system maintains homeostasis whereas the endocrine system does not.

 [B] Endocrine glands produce neurotransmitters whereas nerves produce hormones.

 [C] Nerve signals travel on neurons whereas hormones travel through the blood.

 [D] The nervous system involves chemical transmission whereas the endocrine system does not.

 [E] The nervous system produces physiological responses whereas the endocrine produces behavioral.

The answer is C.

In the human nervous system, neurons carry nerve signals to and from the cell body. Endocrine glands produce hormones that are carried through the body in the bloodstream.

36. **Which process generates the most ATP?**

 [A] Fermentation

 [B] Glycolysis

 [C] The Calvin cycle

 [D] The Krebs cycle

 [E] Chemiosmosis

The answer is E.

The electron transport chain uses electrons to pump hydrogen ions across the mitochondrial membrane. This ion gradient is used to form ATP in a process called chemiosmosis. ATP is generated by the removal of hydrogen ions from NADH and FADH2. This yields 34 ATP molecules.

37. Which of the following is a function of the cardiovascular system?

 [A] Move oxygenated blood around the body

 [B] Oxygenate the blood through gas exchange

 [C] Act as an exocrine system

 [D] Flush toxins out of the body

 [E] Transport signals from the brain

 The answer is A.

 The cardiovascular system moves oxygenated blood around the body via the heart (a pump) and tubes (arteries and veins).

38. Which of the following is not a part of the nervous system?

 [A] Brain

 [B] Spinal cord

 [C] Axons

 [D] Venules

 [E] Cochlea

 The answer is D.

 Venules are part of the circulatory system. The others are part of the nervous system.

39. Organisms need to maintain a constant internal environment to survive. Which of the following is a process by which they achieve this?

[A] Respiration

[B] Reproduction

[C] Depolarization

[D] Repolarization

[E] Thermoregulation

The answer is E.

Thermoregulation is the process that enables an organism to maintain its body temperature. If it is an endothermic organism, it can respond to changes in temperature by sweating or growing more fur. If it is an ectothermic organism, it can move to a warmer or cooler location.

40. Which of the following controls the body's endocrine mechanisms?

[A] Feedback loops

[B] Control molecules

[C] Neurochemicals

[D] Neurotransmitters

[E] Behavioral responses

The answer is A.

The body's mechanisms are controlled by feedback loops.

41. Which gland regulates the calcium in the body?

[A] Thyroid gland

[B] Parathyroid gland

[C] Hypothalamus

[D] Pituitary gland

[E] Pancreas

The answer is B.

The parathyroid glands regulate the calcium levels in the body. They are imbedded within the thyroid gland.

42. Which of the following steroids is not created in the gonads?

[A] Testosterone

[B] Estrogen

[C] Progesterone

[D] ACTH

[E] FSH

The answer is D.

ACTH is not one of the three steroids produced by the gonads. The other three are made by the gonads.

43. **What is the most common neurotransmitter?**

 [A] Epinephrine

 [B] Serotonin

 [C] Acetylcholine

 [D] Norepinephrine

 [E] Oxytocin

 The answer is C.

 The most common neurotransmitter is acetylcholine.

44. **Food is carried through the digestive tract by a series of wave-like contractions. What is this process is called?**

 [A] Peristalsis

 [B] Chyme

 [C] Digestion

 [D] Absorption

 [E] Depolarization

 The answer is A.

 Peristalsis is the process of wave-like contractions that moves food through the digestive tract.

45. Which of the following must muscles pull on in order to initiate movement?

 [A] Skin

 [B] Bones

 [C] Joints

 [D] Ligaments

 [E] Bursa

 The answer is B.

 The muscular system's function is for movement. Skeletal muscles are attached to bones and are responsible for their movement.

46. Hormones are essential to the regulation of reproduction. Which organ is responsible for the release of hormones for sexual maturity?

 [A] Pituitary gland

 [B] Hypothalamus

 [C] Pancreas

 [D] Thyroid gland

 [E] Pineal gland

 The answer is B.

 The hypothalamus begins secreting hormones that help mature the reproductive system and stimulate development of the secondary sex characteristics.

47. What is the type of muscle in the human body that is voluntary?

[A] Cardiac

[B] Sarcomere

[C] Smooth

[D] Skeletal

[E] Actin

The answer is D.

Of all of the above, only skeletal muscles are under voluntary control. Actin, a component of contracting filaments, is found in the skeletal muscles of the human body.

48. The wrist is an example of what kind of joint?

[A] Ball and socket

[B] Pivot

[C] Stationary

[D] Hinge

[E] Gliding

The answer is B.

The wrist joint is a pivot joint that rotates.

49. **What is the waterproofing protein in the skin called?**

 [A] Actin

 [B] Epidermis

 [C] Collagen

 [D] Sebum

 [E] Keratin

 The answer is E.

 The waterproofing protein in the skin is called keratin.

50. **What is small flap of tissue called that covers the trachea when swallowing food?**

 [A] Epiglottis

 [B] Larynx

 [C] Connective tissue

 [D] Villi

 [E] Squamous tissue

 The answer is A.

 When you swallow, a small flap of tissue called the epiglottis covers the trachea. This prevents the chewed food from going down the windpipe into the lungs and channels it down the esophagus.

51. **What is the role of neurotransmitters in nerve action?**

 [A] To turn off the sodium pump

 [B] To turn off the calcium pump

 [C] To send impulses to neurons

 [D] To send impulses around the body

 [E] To send impulses from axon to dendrite

 The answer is C.

 The neurotransmitters carry the signals from one neuron to another across a gap called a synapse.

52. **Which substance breaks down fats?**

 [A] Bile produced in the gall bladder

 [B] Lipase produced in the gall bladder

 [C] Glucagons produced in the liver

 [D] Amylase produces in the gall bladder

 [E] Bile produced in the liver

 The answer is E.

 The liver produces bile, which breaks down and emulsifies fatty acids.

53. **Where does fertilization in humans usually occur?**

 [A] Uterus

 [B] Ovary

 [C] Fallopian tubes

 [D] Vagina

 [E] Epididymis

 The answer is C.

 Fertilization of the egg by the sperm normally occurs in the fallopian tubes. The fertilized egg is then implanted in the uterine lining for development.

54. **Which of the following is lacking in the dermis layer of skin?**

 [A] Sweat glands

 [B] Keratin

 [C] Hair follicles

 [D] Blood vessels

 [E] Living cells

 The answer is B.

 Keratin is a waterproofing protein found in the epidermis.

55. A school age boy had the mumps as a baby. Why will he most likely not get this disease again?

[A] Passive immunity

[B] Vaccination

[C] Antibiotics

[D] Active immunity

[E] Antigen production

The answer is D.

Active immunity develops after recovery from an infectious disease, such as measles, or after vaccination. Passive immunity to some diseases may be passed from one individual to another (from mother to nursing child).

56. What is any foreign particle called that causes an immune reaction?

[A] An antigen

[B] A histocompatibity complex

[C] An antibody

[D] A vaccine

[E] A bacteriophage

The answer is A.

An antigen is any foreign particle that results in an immune reaction, especially the production of antibodies.

57. **Which of the following statements describes the polymerase chain reaction?**

 [A] It is a group of polymerases.

 [B] It is a technique for amplifying DNA.

 [C] It is a primer for DNA synthesis.

 [D] It is a way to synthesize polymerase.

 [E] It is a series of genetic mutations.

 The answer is B.

 PCR is a technique in which a piece of DNA can be amplified into billions of copies within a few hours.

58. **Which part of a DNA nucleotide can vary?**

 [A] Deoxyribose

 [B] Phosphate group

 [C] Hydrogen bonds

 [D] Sugar

 [E] Nitrogenous base

 The answer is E.

 DNA is made of a 5-carbon sugar (deoxyribose), a phosphate group, and a nitrogenous base. There are four nitrogenous bases in DNA that vary to allow for the four different nucleotides.

59. A DNA strand has the base sequence of TCAGTA. Its DNA complement would have which of the following sequences?

[A] ATGACT

[B] TCAGTA

[C] AGUCAU

[D] AGTCAT

[E] TCTGTA

The answer is D.

The complement strand to a single strand DNA molecule has a complementary sequence to the template strand. T pairs with A and C pairs with G. Therefore, the complement to TCAGTA is AGTCAT.

60. Which of the following carries amino acids to the ribosome during protein synthesis?

[A] Messenger RNA

[B] Ribosomal RNA

[C] Transfer RNA

[D] DNA

[E] RNA

The answer is C.

The tRNA molecule is specific for a particular amino acid. The tRNA has an anticodon sequence that is complementary to the codon. This specifies where the tRNA places the amino acid in protein synthesis.

61. A protein is 60 amino acids in length. This requires a coded DNA sequence of how many nucleotides?

 [A] 20

 [B] 30

 [C] 120

 [D] 180

 [E] 240

 The answer is D.

 Each amino acid codon consists of 3 nucleotides. If there are 60 amino acids in a protein, then 60 x 3 = 180 nucleotides.

62. A DNA molecule has the sequence of ACTATG. What is the anticodon of this molecule?

 [A] UGAUAC

 [B] ACUAUG

 [C] TGATAC

 [D] ACTATG

 [E] CTGCGA

 The answer is D.

 The DNA is first transcribed into mRNA. Here, the DNA has the sequence ACTATG; therefore, the complementary mRNA sequence is UGAUAC (remember, in RNA, T is U). This mRNA sequence is the codon. The anticodon is the complement to the codon. The anticodon sequence will be ACUAUG (remember, the anticodon is tRNA, so U is present instead of T).

63. **What is the general term for a change that affects the sequence of bases in a gene?**

 [A] Deletion

 [B] Polyploid

 [C] Mutation

 [D] Duplication

 [E] Substitution

 The answer is C.

 A mutation is an inheritable change in DNA. It may be an error in replication or a spontaneous rearrangement of one ore more segments of DNA. Deletion and duplication are types of mutations. Polyploidy occurs when an organism has more than two complete chromosome sets.

64. **Segments of DNA can be transferred from the DNA of one organism to another through the use of which of the following?**

 [A] Bacterial plasmids

 [B] Viruses

 [C] Chromosomes from frogs

 [D] Plant DNA

 [E] Okazaki fragments

 The answer is A.

 Plasmids can transfer themselves (and therefore their genetic information) by a process called conjugation. This requires cell-to-cell contact.

65. What is the enzyme that unwinds DNA during replication?

 [A] DNAse

 [B] DNA replicase

 [C] DNA helicase

 [D] DNA topoisomerases

 [E] DNA polymerase

 The answer is C.

 The enzyme helicase is involved in unwinding DNA during replication.

66. What is a small circular piece of DNA called that contains accessory DNA?

 [A] Mitochondrial DNA

 [B] Messenger RNA

 [C] Transfer DNA

 [D] Okazaki fragment

 [E] Plasmid

 The answer is E.

 A plasmid is a small, circular piece of accessory DNA often found in bacteria.

67. In DNA, adenine bonds with _____, while cytosine bonds with _____.

[A] Thymine/guanine

[B] Adenine/cytosine

[C] Cytosine/uracil

[D] Guanine/thymine

[E] Uracil/adenine

The answer is A.

In DNA, adenine pairs with thymine and cytosine pairs with guanine because of their nitrogenous base structures.

68. Which protein structure consists of the coils and folds of polypeptide chains?

[A] Secondary structure

[B] Quaternary structure

[C] Tertiary structure

[D] Primary structure

[E] Quinary structure

The answer is A.

The primary structure is the protein's unique sequence of amino acids. The secondary structure is the coils and folds of polypeptide chains. The coils and folds are the result of hydrogen bonds along the polypeptide backbone. The tertiary structure is formed by bonding between the side chains of the amino acids. The quaternary structure is the overall structure of the protein from the aggregation of two or more polypeptide chains.

69. **What can be said about homozygous individuals?**

 [A] They have two different alleles.

 [B] They are of the same species.

 [C] They exhibit the same features.

 [D] They have a pair of identical alleles.

 [E] They produce identical offspring.

 The answer is D.

 Homozygous individuals have a pair of identical alleles while heterozygous individuals have two different alleles.

70. **The term "phenotype" refers to which of the following?**

 [A] A condition that is heterozygous

 [B] The genetic makeup of an individual

 [C] A condition that is homozygous

 [D] The expression of the genotype

 [E] From which parent the traits were inherited

 The answer is D.

 Phenotype is the physical appearance or expression of an organism due to its genetic makeup (genotype).

71. The ratio of brown-eyed to blue-eyed children from the mating of a blue-eyed male to a heterozygous brown-eyed female is expected to be which of the following?

[A] 3:1

[B] 2:2

[C] 1:0

[D] 1:2

[E] 0:4

The answer is B.

Use a Punnett square to determine the ratio.

B = brown eyes, b = blue eyes. Female genotype is on the side and the male genotype is across the top.

	B	b
b	Bb	bb
b	Bb	bb

The female is heterozygous and her phenotype is brown eyes. This means the dominant allele is for brown eyes. The male expresses the homozygous recessive allele for blue eyes. Their children are expected to have a ratio of brown eyes to blue eyes of 2:2; or 1:1.

72. **Which of the following defines the Law of Segregation defined by Gregor Mendel?**

 [A] After meiosis, each new cell will contain an allele that is recessive.

 [B] Only one of two alleles is expressed in a heterozygous organism.

 [C] The allele expressed is always the dominant allele.

 [D] Alleles of one trait do not affect the inheritance of alleles on another chromosome.

 [E] When sex cells form, the two alleles that determine a trait will end up on different gametes.

 The answer is E.
 The Law of Segregation states that the two alleles for each trait segregate into different gametes.

73. **Which of the following is an example of the incomplete dominance that occurs when a white flower is crossed with a red flower?**

 [A] Pink flowers

 [B] Red flowers

 [C] White flowers

 [D] Red and white flowers

 [E] White and pink flowers

 The answer is A.
 Incomplete dominance is when the F1 generation results in an appearance somewhere between the parents. Red flowers crossed with white flowers results in an F1 generation with pink flowers.

74. A child with type O blood has a father with type A blood and a mother with type B blood. The genotypes of the parents respectively would be which of the following?

[A] AA and BO

[B] AO and BO

[C] AA and BB

[D] AO and OO

[E] OO and AB

The answer is B.

Type O blood has 2 recessive O genes. A child receives one allele from each parent; therefore, each parent in this example must have an O allele. The father has type A blood with a genotype of AO and the mother has type B blood with a genotype of BO.

75. Crossing over, which increases genetic diversity, occurs during which stage(s) of meiosis?

[A] Telophase II in meiosis

[B] Metaphase in mitosis

[C] Interphase in meiosis

[D] Prophase I in meiosis

[E] Metaphase II in meiosis

The answer is D.

During prophase I of meiosis, the replicated chromosomes condense and pair with their homologues in a process called synapsis. Crossing over, the exchange of genetic material between homologues to further increase diversity, occurs during prophase I of meiosis.

76. **ABO blood grouping is an example of which type of allele dominance?**

 [A] Autosomal dominance

 [B] Incomplete dominance

 [C] Somatic dominance

 [D] Complete dominance

 [E] Codominance

 The answer is C.

 ABO blood grouping involves codominance. This means that more than one allele can express itself at the same time.

77. **In a Punnett square with a single trait, what are the ratios of genotypes produced between two heterozygous individuals?**

 [A] 1:2:2

 [B] 2:1:1

 [C] 1:1:1

 [D] 1:2:1

 [E] 2:2:2

 The answer is D.

 The Punnett square ratio for a single trait is 1:2:1. All three possible genotypes will be expressed: homozygous dominant, heterozygous, and homozygous recessive.

78. **What is the term for an organism's genetic makeup?**

 [A] Heterozygote

 [B] Genotype

 [C] Phenotype

 [D] Homozygote

 [E] Dominance

 The answer is B.

 The genetic makeup is called the genotype.

79. **Which of the following represents a genetic engineering advancement in the medical field?**

 [A] Stem cell reproduction

 [B] Pesticides

 [C] Degradation of harmful chemicals

 [D] Antibiotics

 [E] Gene therapy

 The answer is E.

 Gene therapy is the introduction of a normal allele to the somatic cells to replace a defective allele. The medical field has had success in treating patients with a single enzyme deficiency disease. Gene therapy has allowed doctors and scientists to introduce a normal allele that provides the missing enzyme.

80. Which of the following is not true regarding restriction enzymes?

 [A] They aid in transcombination procedures.

 [B] They are used in genetic engineering.

 [C] They are named after the bacteria in which they naturally occur.

 [D] They identify and splice certain base sequences on DNA.

 [E] They can be produced by certain lipids during DNA replication.

 The answer is A.

 A restriction enzyme is a bacterial enzyme that cuts foreign DNA at specific locations. The splicing of restriction fragments into a plasmid results in a recombinant plasmid.

81. Which of the following processes is not one of the modern uses of DNA?

 [A] PCR technology

 [B] Gene therapy

 [C] Cloning

 [D] Genetic Alignment

 [E] Transgenic organisms

 The answer is D.

 PCR technology, gene therapy and cloning all come out of working with DNA.

82. **Which statement best represents gel electrophoresis?**

 [A] It isolates fragments of DNA for scientific purposes.

 [B] It cannot be used in proteins.

 [C] It requires the polymerase chain reaction.

 [D] It only separates DNA by size.

 [E] It uses different charged particles to color the bands.

 The answer is A.

 Gel electrophoresis separates DNA by size and charge. It can be used in proteins as well and is not dependent on the polymerase chain reaction.

83. **What is the term that describes the duplication of genetic material in another cell?**

 [A] Replicating

 [B] Cell duplication

 [C] Transgenics

 [D] Genetic restructuring

 [E] Cloning

 The answer is E.

 Cloning is the duplication of genetic material in another cell; the new cell is an exact replica of the original cell.

84. What does gel electrophoresis use to separate the DNA?

[A] The amount of current

[B] The size of the molecule

[C] The positive charge of the molecule

[D] The solubility of the gel

[E] The source of the DNA

The answer is B.

Electrophoresis uses electrical charges of molecules to separate them according to their size.

85. **Which of the following is a result of reproductive isolation?**

[A] Extinction

[B] Migration

[C] Fossilization

[D] Speciation

[E] Radiation

The answer is D.

Reproductive isolation is caused when two groups of organisms can no longer exchange genes. As a result, each group shares traits with only those members with which it can reproduce. Reproductive isolation of populations is the primary criterion for recognition of species status.

86. **Which of the following is true about natural selection?**

 [A] It acts on an individual genotype.

 [B] It is not currently happening.

 [C] It is only an animal phenomenon.

 [D] It acts on the individual phenotype.

 [E] It is used to prevent overpopulation.

 The answer is D.

 Natural selection acts on the individual phenotype and, in so doing, indirectly adapts a population to its environment by maintaining favorable genotypes in the gene pool.

87. **How does diversity aid a population?**

 [A] Individuals are better able to survive.

 [B] Mates are attracted to a diverse population.

 [C] Potential mates like conformity.

 [D] It increases the DNA differences in the population.

 [E] It provides possible improvements to the population.

 The answer is E.

 Diversity provides possible improvements to the population that may help individuals survive and reproduce.

88. **Which statement is not true about diversity?**

 [A] Without diversity there would be extinction.

 [B] Diversity is increasing all the time.

 [C] Fossil evidence supports diversity.

 [D] Sexual reproduction encourages more diversity.

 [E] Skeletons are too similar to allow for diversity.

 The answer is E.

 The other answers are all true. Without diversity, there would be extinction, diversity is increasing all the time and fossil evidence supports an increase in diversity.

89. **Which of the following ideas was a major part of Darwin's evolutionary theory?**

 [A] Punctualism

 [B] Gradualism

 [C] Equilibrium

 [D] Convergency

 [E] Altruism

 The answer is B.

 Darwin's book is based upon gradualism, the idea that species change slowly over time.

90. **Which statement is not true about reproductive isolation?**

 [A] It prevents populations from exchanging genes.

 [B] It can occur by preventing fertilization.

 [C] It can result in speciation.

 [D] It happens more often on the mainland.

 [E] It produces offspring with unique phenotypes.

 The answer is D.

 Reproductive isolation can result in speciation, can occur by preventing fertilization and prevents populations from exchanging genes. It is a common phenomenon on islands.

91. **Which idea is true about members of the same species?**

 [A] They look identical.

 [B] They never change.

 [C] They reproduce successfully within their group.

 [D] They live in the same geographic location.

 [E] They have very dissimilar genotypes.

 The answer is C.

 Species are defined by the ability to successfully reproduce with members of their own kind.

92. Which of the following factors will affect the Hardy-Weinberg law of equilibrium, leading to evolutionary change?

[A] No mutations

[B] Non-random mating

[C] No immigration or emigration

[D] Large population

[E] Small individual species

The answer is B.

There are five requirements to keep the Hardy-Weinberg equilibrium stable: no mutation, no selection pressures, an isolated population, a large population, and random mating.

93. If a population is in Hardy-Weinberg equilibrium and the frequency of the recessive allele is 0.3, what percentage of the population is expected to be heterozygous?

[A] 9%

[B] 49%

[C] 42%

[D] 21%

[E] 7%

The answer is C.

0.3 is the value of q. Therefore, $q^2 = 0.09$. According to the Hardy-Weinberg equation, $1 = p + q$.

$1 = p + 0.3$
$p = 0.7$
$p^2 = 0.49$

Next, plug q2 and p2 into the equation $1 = p^2 + 2pq + q^2$.

$1 = 0.49 + 2pq + 0.09$ (where 2pq is the number of heterozygotes).
$1 = 0.58 + 2pq$
$2pq = 0.42$

Multiply by 100 to get the percent of heterozygotes, 42%.

94. Which aspect of science does not support evolution?

 [A] Comparative anatomy

 [B] Organic chemistry

 [C] Comparison of DNA among organisms

 [D] Analogous structures

 [E] Embryology

The answer is B.

Comparative anatomy is the comparison of characteristics of the anatomies of different species. This includes homologous structures and analogous structures. The comparison of DNA between species is the best known way to place species on the evolution tree. Embryology indicates common origins among organisms in the early stages of development. Organic chemistry has nothing to do with evolution.

95. In which of the following does evolution occur?

 [A] Individuals

 [B] Populations

 [C] Organ systems

 [D] Cells

 [E] Ecosystems

The answer is B.

Evolution is a change in genotype over time. Gene frequencies shift and change from generation to generation. Populations evolve, not individuals.

96. **Which process contributes most to the large variety of living things in the world today?**

 [A] Meiosis

 [B] Asexual reproduction

 [C] Mitosis

 [D] Alternation of generations

 [E] Reproductive isolation

 The answer is A.

 During meiosis prophase I crossing over occurs. This exchange of genetic material between homologues increases diversity.

97. **Which of the following gases was a major part of the primitive Earth's atmosphere?**

 [A] Fluorine

 [B] Methane

 [C] Oxygen

 [D] Krypton

 [E] Argon

 The answer is B.

 The primitive atmosphere contained ammonia, methane and hydrogen but very little oxygen.

98. **What is a major principle of the Endosymbiotic Theory?**

 [A] Birds and dinosaurs share a common ancestor.

 [B] Animals evolved in close relationships with one another.

 [C] Prokaryotes arose from eukaryotes.

 [D] Inorganic compounds are the basis of living things.

 [E] Eukaryotes arose from very simple prokaryotes.

 The answer is E.

 The Endosymbiotic theory of the origin of eukaryotes states that eukaryotes arose from symbiotic groups of prokaryotic cells. According to this theory, smaller prokaryotes lived within larger prokaryotic cells, eventually evolving into chloroplasts and mitochondria.

99. **The wing of a bird, the human arm, and the pectoral fin of a whale all have the same bone structure. What are these structures called?**

 [A] Polymorphic structures

 [B] Homologous structures

 [C] Vestigial structures

 [D] Analogous structures

 [E] Allopatric structures

 The answer is B.

 Homologous structures have the same genetic basis (leading to similar appearances), but are used for different functions.

100. Which of the following is not an abiotic factor?

[A] Temperature

[B] Rainfall

[C] Soil quality

[D] Predation

[E] Wind speed

The answer is D.

Abiotic factors are non-living aspects of an ecosystem. Temperature, rainfall, and soil quality are all abiotic factors. Predation is an example of a biotic factor—living things acting on each other.

101. Which of the following is not true about cladistics?

[A] It is the study of phylogenetic relationships of organisms.

[B] It involves a branching diagram that uses the development of novel traits to separate groups of organisms.

[C] It distinguishes between the relative importance of the traits.

[D] It shows when traits developed with respect to other traits.

[E] It indicates which organisms are most closely related to each other and what their common ancestors were.

The answer is C.

Cladistics does not show how important certain traits were to different species. It represents when species evolved and how closely related they are to each other.

102. If DDT were present in an ecosystem, which of the following organisms would have the highest concentration in its body?

[A] Herring

[B] Diatom

[C] Zooplankton

[D] Salmon

[E] Osprey

The answer is E.

Chemicals and pesticides accumulate along the food chain. Tertiary consumers have more accumulated toxins than animals at the bottom of the food chain.

103. What eats secondary consumers?

[A] Producers

[B] Tertiary consumers

[C] Primary consumers

[D] Decomposers

[E] Detritivores

The answer is B.

The tertiary consumers eat the secondary consumers and the secondary consumers eat the primary consumers.

104. **Which statement is true about the water cycle?**

 [A] Two percent of the water is fixed and unavailable.

 [B] 75% of available water is groundwater.

 [C] The water cycle is driven by the ocean currents.

 [D] Surface water percolates up from underground springs.

 [E] New water is being added into the cycle all the time.

 The answer is A.

 96 percent of available water is groundwater. The water cycle is driven by the sun. Surface water is available.

105. **Which statement about the carbon cycle is false?**

 [A] Ten percent of all available carbon is in the air.

 [B] Carbon dioxide is fixed by glycosylation.

 [C] Plants fix carbon in the form of glucose.

 [D] Animals release carbon through respiration.

 [E] Most atmospheric carbon comes from the decay of dead organisms.

 The answer is B.

 Ten percent of all available carbon is in the air. Plants fix carbon via photosynthesis to make glucose and animals release carbon through respiration.

106. What is the impact of sulfur oxides and nitrogen oxides in the environment when they react with water?

 [A] Ammonia

 [B] Acidic precipitation

 [C] Sulfuric acid

 [D] Global warming

 [E] Greenhouse effect

 The answer is B.

 Acidic precipitation is rain, snow, or fog with a pH less than 5.6. It is caused by sulfur oxides and nitrogen oxides that react with water in the air to form acids that fall to Earth as precipitation.

107. Which term is not associated with the water cycle?

 [A] Precipitation

 [B] Transpiration

 [C] Fixation

 [D] Evaporation

 [E] Runoff

 The answer is C.

 Water is recycled through the processes of evaporation and precipitation. Transpiration is the evaporation of water from leaves. Fixation is not associated with the water cycle.

108. Which of the following is a density dependent factor that affects a population?

[A] Temperature

[B] Rainfall

[C] Predation

[D] Soil nutrients

[E] Wind speed

The answer is C.

As a population increases, the competition for resources is intense and the growth rate declines. This is a density-dependent factor. An example of this would be competition. Density-independent factors affect the population regardless of its size. Examples of density-independent factors are rainfall, temperature, and soil nutrients.

109. High humidity and temperature stability are present in which of the following biomes?

[A] Taiga

[B] Deciduous forest

[C] Desert

[D] Tropical rain forest

[E] Coniferous forest

The answer is D.

A tropical rain forest is located near the equator. Its temperature ranges around 25 degrees C and the humidity is high due to the rainfall that exceeds 200 cm per year.

110. Which trophic level is the most ecologically efficient?

[A] Decomposers

[B] Producers

[C] Tertiary consumers

[D] Secondary consumers

[E] Primary consumers

The answer is B.

The amount of energy that is transferred between trophic levels is called the ecological efficiency. The visual of this is represented in a pyramid of productivity. The producers have the greatest amount of energy and are at the bottom of this pyramid.

111. From where does the oxygen created in photosynthesis come?

[A] Carbon dioxide

[B] Chlorophyll

[C] Glucose

[D] Carbon monoxide

[E] Water

The answer is E.

In photosynthesis, water is split; the hydrogen atoms are pulled to carbon dioxide that is taken in by the plant and ultimately reduced to make glucose. The oxygen from the water is given off as a waste product.

112. Which of the following is true of decomposers?

[A] Decomposers recycle the carbon accumulated in durable organic material.

[B] Decomposers take nitrogen out of the soil to use for food.

[C] Decomposers absorb nutrients from the air to maintain their metabolisms.

[D] Decomposers belong to the Genus Escherichia.

[E] Decomposers are able to use the sun to produce their own energy.

The answer is A.

Decomposers recycle phosphorus and carbon and undergo ammonification. They break down dead organisms to release the carbon held within their tissues. This carbon then reenters the ecosystem.

113. A clownfish is protected by a sea anemone's tentacles, and in turn, the anemone receives uneaten food from the clownfish. What type of relationship is exemplified by this example?

[A] Mutualism

[B] Parasitism

[C] Commensalism

[D] Competition

[E] Amensalism

The answer is A.

Neither the clownfish nor the anemone cause harmful effects towards one another and they both benefit from their relationship. Mutualism occurs when two species that occupy a similar space benefit from their relationship.

114. Which of the following is most likely to happen in order for primary succession to occur?

 [A] Nutrient enrichment

 [B] A forest fire

 [C] Bare rock is exposed after a water table recedes

 [D] A housing development is built

 [E] A farmer stops cultivating her fields

 The answer is C.

 Primary succession occurs where life never existed before, such as previously flooded areas or a new volcanic island. It is only after the water recedes that the rock is able to support new life.

115. What is the Mendelian law called that states that only one of the two possible alleles from each parent is passed on to the offspring?

 [A] The Mendelian Law

 [B] The Law of Independent Assortment

 [C] The Law of Segregation

 [D] The Allele Law

 [E] The Law of Dominance and Recessiveness

 The answer is B.

 The law of independent assortment states that only one of a pair of alleles is transferred from parent to offspring.

CLEP Biology Sample Exam 2

None of the following sample questions has appeared on an actual CLEP examination. They are intended to show you the types of questions and level of difficulty that you will encounter on the actual examination. The exam will indicate topics you may need to study further, and the questions themselves will provide you with content for practice and review. Keep in mind that knowing the answers to all of the sample questions does not a guarantee that you will pass the CLEP Biology exam.

Directions: Each question has five possible answers. For each, choose the one that you feel best answers the question or completes the statement.

1. **What is the main function of enzymes?**

 [A] They lower the activation energies of chemical reactions, thereby speeding them up.

 [B] They activate certain hormones within bacterial cells to decrease their rate of binary fission.

 [C] They increase of net gain of ATP produced during cellular respiration in order to improve cell function.

 [D] They convert light energy into chemical energy during photosynthesis so a plant can then produce sugars.

 [E] They remove urea from the excretory system so toxicity levels within the organism do not reach dangerous levels.

2. **What is the term given to new species colonizing an area over time after a natural disaster?**

 [A] Secondary succession

 [B] Interspecific competition

 [C] Intraspecific competition

 [D] Primary succession

 [E] Tertiary succession

3. What kind of growth curve is being shown in the graph for a population of bacteria in a Petri dish?

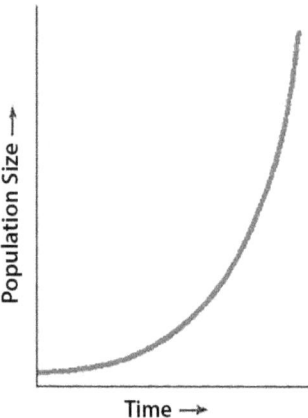

[A] Logistical growth

[B] Exponential growth

[C] Preferential growth

[D] Vertical growth

[E] Redundancy growth

4. Over a 20-year span of time, a female whale produced all male offspring. She is pregnant again. What is the probability that this next offspring will be male?

[A] 10%

[B] 50%

[C] 60%

[D] 75%

[E] 100%

Questions 5 and 6 refer to the following information:

Achondroplasia is a dominant genetic trait that causes dwarfism (stunted growth). The homozygous condition is lethal. Heterozygotes, however, express the dwarf trait. Rupert has a family history of dwarfism, but does not express it himself. His wife, Lola, has achondroplasia.

5. What are the respective genotypes of Rupert and Lola?

 [A] Aa, aa

 [B] aa, Aa

 [C] Aa, AA

 [D] aa, AA

 [E] Aa, Aa

6. What is the probability their child will have achondroplasia?

 [A] 0 %

 [B] 25%

 [C] 50%

 [D] 75%

 [E] 100%

7. What type of behavior is a bean plant exhibiting when it bends towards the light?

 [A] Positive photokinesis

 [B] Negative phototaxis

 [C] Positive phototaxis

 [D] Negative gravitropism

 [E] Positive phototropism

8. A person with type AB blood can receive a transfusion from a person with which blood type(s)?

[A] Type A only

[B] Type A and B

[C] Type B only

[D] Type A, B, and O

[E] Type O only

Questions 9 and 10 relate to the data table below:

Oxygen Production	
Distance From Light (cm)	Bubbles Produced per Minute
10	39
20	22
30	8
40	5

9. Based upon the finding from this study, what conclusion can be drawn about why primary producers have to live within the uppermost regions of the oceans, ponds, and lakes?

[A] They have to live close to the surface in order to hunt for food.

[B] They have to live close to the surface in order to obtain the light they need for photosynthesis.

[C] They have to live close to the surface in order to produce enough energy to reproduce.

[D] They have to live close to the surface in order to excrete toxic levels of carbon dioxide.

[E] They have to live close to the surface in order to maximize their ability to absorb oxygen from the air.

10. Which is the correct process being supported by the data?

[A] $H_2O + CO_2 \xrightarrow{light} sugars + O_2$

[B] $sugars + O_2 \xrightarrow{light} H_2O + CO_2$

[C] $H_2O + O_2 \xrightarrow{light} sugars + CO_2$

[D] $sugars + CO_2 \xrightarrow{light} H_2O + O_2$

[E] $O_2 + CO_2 \xrightarrow{light} CO_2 + sugars$

11. The wing of a bat, the human hand, the fin of a whale, and the front leg of a dog all share a similar bone structure. How do these traits support the theory of evolution?

[A] They show diversity between species.

[B] They show a common ancestor among all species.

[C] They show how Lamarckism can pass traits between species.

[D] They show the existence of a unified anatomical theme of all organisms.

[E] They show how adaptive radiation can lead to the formation of new traits.

12. Which of the following represents the most likely timeline of the evolution of life on Earth, starting with the oldest?

[A] Plants → fungi → animals → humans

[B] Fungi → bacteria → plants → protists

[C] Bacteria → protists → plants → animals

[D] Protists → humans → animals → plants

[E] Humans → animals → plants → fungi

13. Which of the following correctly lists the embryological stages of development in animals?

 [A] Zygote, blastula, gastrula, morula

 [B] Zygote, blastula, morula, gastrula

 [C] Zygote, gastrula, blastula, morula

 [D] Zygote, morula, blastula, gastrula

 [E] Zygote, morula, gastrula, blastula

14. Monarch butterflies have an orange and black coloration that identifies them well known to predators. Since these butterflies eat milkweed, which is poisonous to other creatures, the predators know to stay away from them. The viceroy butterfly also has an orange and black coloration that is very similar to the monarch's, but the viceroy butterfly does not feed on milkweed, so it is not toxic. Even so, predators still avoid the viceroy. What is the adaptation called that keeps predators from eating the viceroy butterflies?

 [A] Mutation

 [B] Learning

 [C] Reproductive isolation

 [D] Pattern formation

 [E] Mimicry

15. If one parent has the genotype AABBCCDDEE and the other parent has the genotype aabbccddee, what are the possible genotypes of their offspring?

 [A] All AaBbCcDdEe

 [B] AABBccDDEE and aaBBCCDDEE

 [C] aaBBccDDee and AAbbCCddEE

 [D] AaBBCcDDEE and aABbCCDdEE

 [E] AAbbCCddee and AAbbCCddee

Number 16–19. Use the following options to answer the questions.

 [A] Lipids

 [B] Proteins

 [C] Carbohydrates

 [D] Nucleic Acids

16. Which macromolecule has monomers called amino acids?

17. Which macromolecule does the body use to store energy for the long term?

18. Which of these contains instructions for all of the cell's activities?

19. Which macromolecule is good for giving a quick burst of energy?

20. Which of the following do NOT exhibit meiotic cell division?

 [A] Mushroom

 [B] Tree

 [C] Dog

 [D] Bacteria

 [E] Amoeba

21. Which of these refers to the random distribution of maternal and paternal chromosomes into daughter cells?

 [A] Translocation

 [B] Crossing-over

 [C] Non-disjunction

 [D] Independent assortment

 [E] Dominance and recessiveness

22. Which of these organisms is most closely related to the one with the scientific name *Felis concolor*?

 [A] *Canis lupis*

 [B] *Paris bicolor*

 [C] *Felis domesticus*

 [D] *Paris atricapulis*

 [E] *Callinectes sapidus*

23. What characteristic is the distinguishing feature of birds?

 [A] Birds are able to lay eggs.

 [B] Birds have bodies covered with feathers.

 [C] Birds migrate to warmer climates during the winter.

 [D] Birds build complex nests to protect their young.

 [E] Birds eat many different types of foods for energy.

24. In an experiment, lab mice had their tails snipped soon after birth for 15 generations. When the mice were reproduced for a 16th time, all of the offspring had tails. What conclusion can be drawn from these results?

[A] The mice were showing natural selection.

[B] The mice were exhibiting the principles of Lamarckism.

[C] The mice showed little or no similarities to their parents.

[D] The mice showed that mutations in gametes could affect offspring traits.

[E] The mice showed that offspring inherit characteristics acquired by their parents.

Questions 25 – 29 relate to the following organs found within mammals.

[A] Stomach

[B] Small intestine

[C] Liver

[D] Pancreas

[E] Large intestine

25. Which organ creates the enzyme that breaks down fats?

26. Which organ is the site of most chemical digestion?

27. Which organ is responsible for the absorption of water from the waste products of digestion?

28. Which organ breaks apart and grinds up food that enters it from the mouth?

29. Which organ secretes the enzyme that is mainly responsible for the digestion of proteins?

30. Which of the following is true regarding the flow of energy and nutrients through an ecosystem?

 [A] Energy and nutrients are both recycled.

 [B] Both energy and nutrients accumulate in the highest trophic levels.

 [C] Only 10 % of the available energy is recycled due to loss from metabolism.

 [D] Nutrients get recycled, but most of the energy is lost from every trophic level.

 [E] The amount of nutrients available for each tropic level is depends on the available energy.

31. Which of the following BEST explains why there are usually fewer than five trophic levels to most food chains?

 [A] Many primary consumers feed at more than one trophic level.

 [B] The carrying capacity of the environment would be exceeded with more than five levels.

 [C] Ecosystems with more than five levels contain too much biomass.

 [D] The increased demand on the tertiary consumers would cause them to face extinction.

 [E] Each trophic level only obtains a small fraction of the energy from the trophic level below it.

Questions 32–34 refer to the following information:

Humans can have either attached earlobes (f) or free earlobes (F). Assume that two parents are heterozygous for free earlobes.

32. What are the genotypes of the parents?

 [A] Ff x Ff

 [B] Ff x ff

 [C] ff x Ff

 [D] FF x ff

 [E] ff x FF

33. What percentage of the parent's sex cells carried the allele for free earlobes?

 [A] 0%

 [B] 25%

 [C] 50%

 [D] 75%

 [E] 100%

34. What are the possible phenotypes of the offspring?

 [A] All offspring will have free earlobes.

 [B] All offspring will have attached earlobes.

 [C] 25% of the offspring will have free earlobes.

 [D] 75% of the offspring will have free earlobes.

 [E] 100% of the offspring will have free earlobes.

35. Which of the following is an example of a polysaccharide?

 [A] Saturated fats

 [B] Cholesterol

 [C] Glucose

 [D] Starch

 [E] Lysine

36. What would most likely happen to a plant cell that was placed into an isotonic solution?

 [A] It would become turgid.

 [B] It would become flaccid.

 [C] It would swell and lyse.

 [D] It would elongate.

 [E] It would undergo apoptosis.

37. In snapdragons, red flowers are dominant to white. A researcher did the following cross and got the reported results.

 red snapdragon x white snapdragon = pink snapdragon

 pink snapdragon x pink snapdragon = red, white, and pink snapdragon

 What is the possible explanation for these results?

 [A] Snapdragons show a condition called codominance, where both alleles show up at the same time.

 [B] Snapdragons demonstrate incomplete dominance, which causes phenotypes to mix in heterozygotes.

 [C] Snapdragons can express different colors, which can be used as indicators of environmental conditions.

 [D] Snapdragons are prone to mutation, so when pink flowers were crossed the different colors were expressed.

 [E] Snapdragons are all heterozygous, which leads to the expression of the dominant and recessive traits when crossed.

38. After competing in a long race, a runner continues to breathe hard even after crossing the finish line. What purpose does this serve?

 [A] It causes the brain to return to a normal resting state.

 [B] It helps the runner to produce more glucose for energy.

 [C] It repays the oxygen debt created from muscle exertion.

 [D] It distributes energy around the body to the runner's organs.

 [E] It drops the runner's blood pressure so the heart rate returns to normal

39. Which of the following predicts what would happen if a participant in a study were given air to breathe that had a higher than usual carbon dioxide concentration?

 [A] They would have increased respiration and heart rate

 [B] They would have decreased respiration and blood pH.

 [C] They would have decreased respiration and heart rate.

 [D] They would have increased respiration and decreased blood pH.

 [E] They would have decreased respiration and increased blood pH.

40. In terms of evolution, which type of organism would be considered the most successful?

 [A] An organism that has the largest territory.

 [B] An organism that eats the greatest variety of food.

 [C] An organism that can reproduce the most often.

 [D] An organism that has the largest amount of biomass.

 [E] An organism that leaves behind the greatest number of offspring.

41. What is the structure called that plants use to protect their leaves from water loss?

 [A] Cuticle

 [B] Dermis

 [C] Parenchyma

 [D] Phloem

 [E] Xylem

42. During the fall, many deciduous trees change color. For example, sugar maples of New England tend to go from green to bright oranges, yellows, and reds. What is the most likely explanation for this?

 [A] The absorption spectrum of chlorophyll changes during the fall to include green wavelengths.

 [B] The trees increase their production of these pigments to adapt to the different amounts of light.

 [C] There is a reduction in the production of green chlorophyll, so the masked pigments become visible.

 [D] The light available at this time of year lacks the blue and green wavelengths found in the summer light.

 [E] There are more orange, yellow, and red wavelengths found in the light during the fall than in the spring and summer.

Questions 43–46 refer to the following choices:

 [A] Mutualism

 [B] Commensalism

 [C] Parasitism

 [D] Predator/prey

 [E] Competition

43. What happens when two species live together and one of them benefits but the other is unharmed?

44. What type of relationship exists when one species hunts and eats another?

45. What type of relationship do a clownfish and a sea anemone exhibit?

46. What type of relationship exists between a tick and a deer?

47. Protein synthesis always results in the production of which of these?

 [A] Ammonia

 [B] ATP

 [C] Water

 [D] Carbon dioxide

 [E] Oxygen

48. What is the mass of chewed food and saliva called that enters the esophagus?

 [A] Bolus

 [B] Duodenum

 [C] Ilium

 [D] Retina

 [E] Trypsin

49. Which of the following is correctly matched with its structure of excretion?

 [A] Fish – nephridia

 [B] Honeybee – flame bulbs

 [C] Planaria – kidneys

 [D] Grasshopper – malphigian tubules

 [E] Humans – spleen

The phylogenetic tree below traces the evolution of plants. Questions 50 – 54 relate to the phylogenetic tree found below.

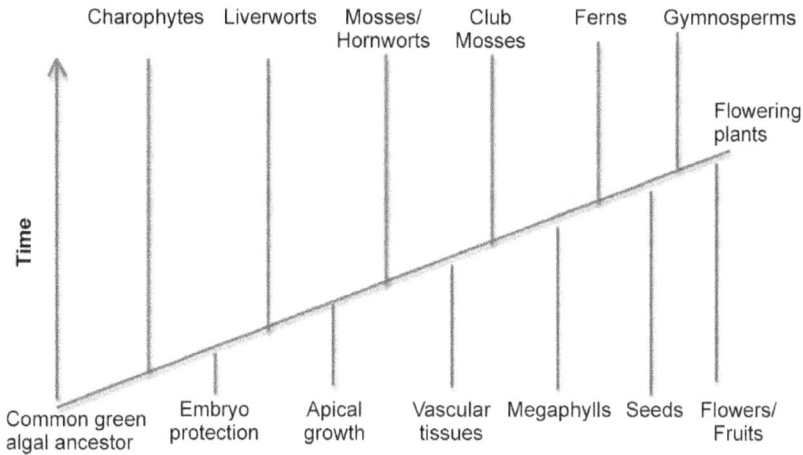

50. Which trait was the earliest to appear in the evolution of plants?

 [A] Seeds

 [B] Flowers

 [C] Apical growth

 [D] Vascular tissues

 [E] Embryo protection

51. Which two types of plants are most closely related to each other?

 [A] Hornworts and ferns

 [B] Flowering plants and gymnosperms

 [C] Charophytes and club mosses

 [D] Liverworts and flowering plants

 [E] Mosses and charophytes

52. **Which plants never evolved the ability to have apical growth?**

 [A] Ferns

 [B] Charophytes

 [C] Hornworts

 [D] Gymnosperms

 [E] Flowering plants

53. **Which structure has led to the success of the most recent land plants?**

 [A] Fruits

 [B] Seeds

 [C] Megaphylls

 [D] Vascular tissue

 [E] Embryo protection

54. **What conclusion can be drawn from the information presented in this phylogenetic tree?**

 [A] The most ancient form of land plants is the moss.

 [B] Vascular tissues were essential to the success of all land plants.

 [C] Ferns are the most closely related land plants to the ancient green algal ancestor.

 [D] Liverworts have adapted and survived without the development of seeds or vascular tissue.

 [E] The progression of time has caused some land plants to become less complex than their ancestors.

55. Which of these animals would be considered endothermic?

 [A] Bacterium

 [B] Crocodile

 [C] Grasshopper

 [D] Hummingbird

 [E] Rattlesnake

Questions 56–59 refer to the food web in the diagram

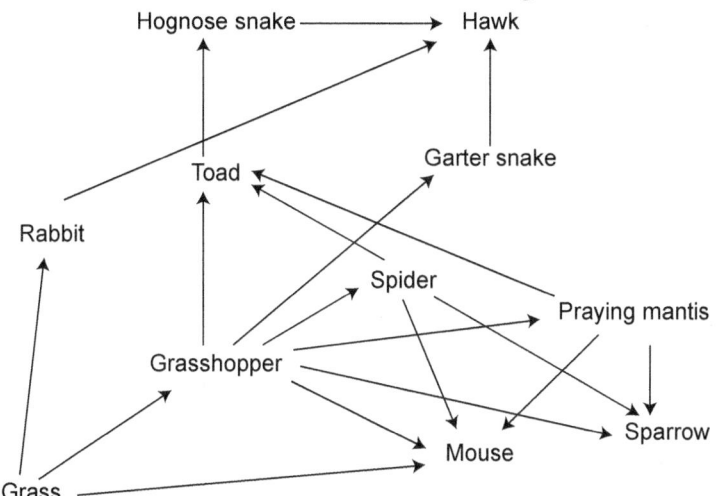

56. Which organisms in this food web would be considered primary consumers?

 [A] Hognose snake, hawk, sparrow

 [B] Spider, rabbit, praying mantis

 [C] Rabbit, grasshopper, mouse

 [D] Grasshopper, toad, hark

 [E] Garter snake, praying mantis, mouse

57. In which sequence of events is the mouse considered a tertiary consumer?

 [A] Grass → grasshopper → spider → mouse

 [B] Grass → rabbit → hawk → mouse

 [C] Grass → mouse → hognose snake → hawk

 [D] Grass → sparrow → praying mantis → garter snake

 [E] Grass → toad → hawk → grasshopper

58. In this food web, what would be a likely impact if the grasshopper population were to decrease?

 [A] The number of rabbits would increase.

 [B] The hawk population would decrease as well.

 [C] The hognose snake population would increase.

 [D] The mouse population would show a steady decline.

 [E] The number of garter snakes would drastically decrease.

59. Why is the toad considered an opportunistic feeder in this food web?

 [A] The toad gets eaten by several different organisms.

 [B] The toad is part of three different food chains.

 [C] The toad's main predator is the hognose snake.

 [D] The toad's main source of food is the hawk's leftovers.

 [E] The toad can survive under many different environmental conditions.

60. Which of the following can feed at more than one level of the trophic pyramid?

 [A] Primary producer

 [B] Secondary consumer

 [C] Omnivore

 [D] Carnivore

 [E] Herbivore

61. Which of the following would be considered a density-independent factor that can cause a change within a population?

 [A] Predation

 [B] Mate selection

 [C] Competition

 [D] Natural disasters

 [E] Niche selection

62. Tapeworms live within the intestines of large mammals where they absorb the nutrients intended for the mammal and use them for their own growth and reproduction. Since the large mammal is not getting any nutrition, it starts to lose weight and wither away. What kind of relationship exists between the tapeworm and the mammal?

 [A] Parasitism

 [B] Mutualism

 [C] Commensalism

 [D] Predator/prey

 [E] Ammensalism

Questions 63 – 67 pertain to the different types of biomes presented below.

[A] Savanna

[B] Desert

[C] Tropical rainforest

[D] Temperate forest

[E] Open ocean

63. Which biome has the richest species diversity?

64. Which biome receives the least amount of rainfall every year?

65. Where would one go to see how light penetration could affect the rates of photosynthesis?

66. Which biome has low shrubs, tall grasses, and herds of animals?

67. Where would one go to have a mild summer, a cold winter, and see the leaves fall off the trees?

68. The fossil record can be used to estimate the total number of taxonomic families that have existed over time. These estimates are shown in the graph. The letters A-D represent periods of mass extinctions. What can be concluded from the data?

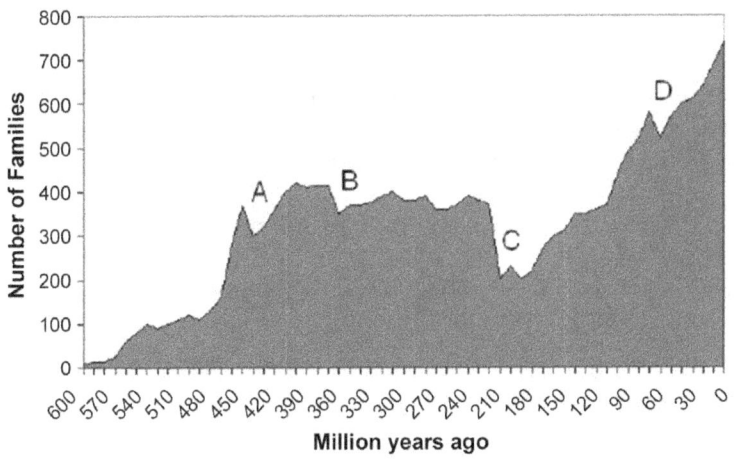

[A] Changes in global climate can result in a mass extinction.

[B] The total amount of biodiversity has difficulty recovering after a mass extinction.

[C] Mass extinctions generally eliminate all life on the planet, so ecosystems need to start over.

[D] Mass extinctions are often followed by periods of speciation and increased biodiversity.

[E] There is incomplete evidence from the fossil record to associate mass extinctions with the biodiversity.

69. While examining a population of fish in a river ecosystem, you notice that the density of healthy adults has increased to a level much higher than your previous record. What can be concluded from this observation?

[A] There was a reduction in the death rate.

[B] There was a reduction in the birth rate.

[C] There was an increase in the rate of immigration.

[D] There was an increase in the rate of emigration.

[E] There was a reduction in the rate of emigration.

70. Which of these would be considered a niche of a beetle living in the tropical rainforest?

[A] The rainforest floor

[B] Living in the canopy

[C] The presence of predators

[D] All of the leaves on a tree

[E] A particular tree in the forest

71. A volcanic eruption has just ended, wiping out all of the trees, grasses, and shrubs on a new island. All of the animals are gone as well. What are the first organisms called that will repopulate this area over time?

[A] Pioneer species

[B] Climax species

[C] Primary producers

[D] Keystone species

[E] Primary consumers

Use the graph of the Age Structure of the World Population to answer questions 72 and 73.

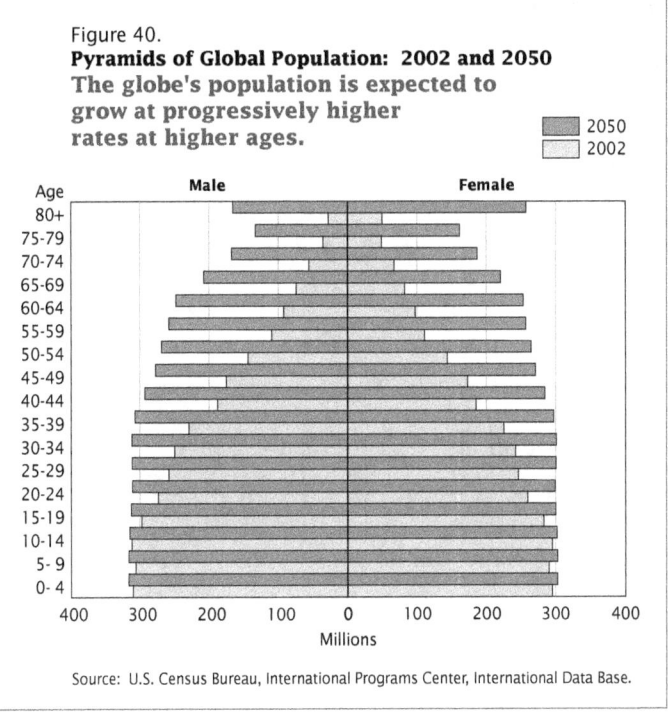

72. Which of these is a trend in the human population between 2002 and 2050?

[A] People are expected to live longer in the future.

[B] There will be consistently more females than males.

[C] Middle age will be found between 55 -59 in 2050.

[D] Children will continue having difficulty living past age 15.

[E] Far more people were alive in 2002 than will be in 2050.

73. **Due to the predicted age structure of the human population, which of the following is likely to eventually happen to the plant's demographics?**

 [A] The number of children below 19 will show a decrease.

 [B] There will be an increase in the number of people over 70 years old.

 [C] There will be a decrease in the number of people between 30 and 50 years of age.

 [D] The overall population of the world will decrease because many will be past reproductive age.

 [E] The populations of some countries will continue to increase at dramatic rates due to lack of preventative measures.

74. **Which of these would be considered a renewable resource?**

 [A] Coal mining

 [B] Oil production

 [C] Solar energy panels

 [D] Natural gas extraction

 [E] Uranium in radioactive decay

75. The zebra mussel (*Dreissena polymorpha*) is a small bivalve that has been known to cause problems for boaters in the Great Lakes by attaching to their propellers and clogging up irrigation pipes, preventing water flow. These animals have few natural predators and are not native to the area. What is the term that describes this type of non-native organism?

 [A] Biodiversity

 [B] Predator

 [C] Invasive species

 [D] Parasitic species

 [E] Polygenic species

76. Which of these is believed to be the greatest cause of the greenhouse effect?

 [A] The melting of the polar ice caps

 [B] The recycling of paper and plastics

 [C] The release of carbon dioxide from factories and cars

 [D] The rise in ocean levels

 [E] The decrease in biodiversity in the tropical rainforests

77. All ecosystems are made up of biotic and abiotic factors. Which statement reflects just the biotic aspects?

 [A] Salmon can live in both fresh and salt water.

 [B] Pandas can only live in areas where bamboo grasses grow.

 [C] Whales have low metabolic rates compared to other mammals

 [D] The seeds of some pine trees need fire in order to germinate.

 [E] Brown algae prefer grow in areas where the water is very cold.

78. A farmer's corn crops are being plagued by a new pest species. These insects eat the leaves and all the seeds on the cobs. She wants to decrease the population of these pests in the most ecologically friendly way possible. What should the farmer do?

[A] The farmer should add pesticides to the water used on the crops.

[B] The farmer should introduce a pathogen into the soil to kill the pests.

[C] The farmer should introduce a natural predator of the pest into the field.

[D] The farmer should add additional competitors for the corn into the system.

[E] The farmer should decrease the carrying capacity of the insect pest's habitat.

Use the information in the graph to answer questions 79 and 80.

In a classic experiment to show competition, two species of paramecium are cultured separately in different containers and then again together in the same container. The population sizes of each species are recoded in the graphs below.

79. What conclusion can be made from examining the population graphs of these two species?

[A] *Paramecium auralia* outcompetes *P. caudatum* when grown together.

[B] *Paramecium caudatum* is a better competitor than *Paramecium auralia*.

[C] Both species are able to survive equally well when grown in the same culture.

[D] If cultured together, the species of Paramecium will interbreed to create a more suitable hybrid.

[E] The resources needed for both species to live successfully are in high supply when cultured together.

80. **What ecological principle is at hand when both species are cultured in the same container?**

 [A] Intraspecific competition

 [B] Predator/prey relationship

 [C] Competitive exclusion

 [D] Mutualistic relationship

 [E] Convergent evolution

Questions 81–85 refer to the diagram below.

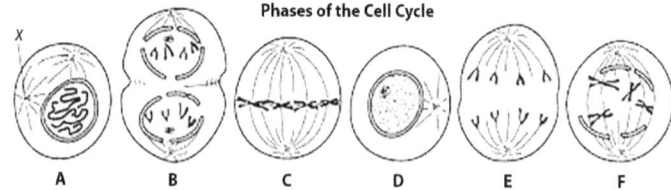

81. **Which represents the cell cycle in the proper order?**

 [A] D, A, F, C, E, B

 [B] A, B, C, D, E, F

 [C] F, E, D, C, B, A

 [D] C, E, B, F, A, D

 [E] B, F, E, A, D, C

82. **Diagram C shows the chromosomes lined up along the equator. During which phase of the cell cycle does this happen?**

 [A] Anaphase

 [B] Cytokinesis

 [C] Metaphase

 [D] Prophase

 [E] Telophase

83. **What is happening during the phase of the cycle represented by letter D?**

 [A] This is cytokinesis where the cell membrane is being split in half.

 [B] This is prophase where the nuclear envelope is being dissolved.

 [C] This is metaphase where the chromosomes line up along the equator of the cell.

 [D] This is anaphase where the individual chromatids get pulled towards opposite poles.

 [E] This is interphase where the genetic material and all the organelles are being duplicated.

84. **What is the function of the process shown in the diagram?**

 [A] To breakdown worn out cell parts

 [B] To enable the organism to grow in size

 [C] To encourage development of new traits

 [D] To destroy foreign invaders within the cell

 [E] To create new genetic combinations that cause evolution

85. **What happens after the cell completes the phase of the diagram shown in B?**

 [A] The cell will sit and rest for a while.

 [B] The cell will jump back into metaphase.

 [C] The cell will divide three more times.

 [D] The cell will enter back into interphase.

 [E] The cell will replace the missing chromatids.

86. What does mitosis in a plant cell have to deal with that mitosis in an animal cell does not?

[A] A cell wall

[B] A nucleus

[C] A nucleolus

[D] A cell membrane

[E] A golgi apparatus

87. What would the complementary strand be to a piece of DNA with the nucleotide sequence AGGTCCGATCA?

[A] AGGTCCGTCA

[B] GAACTTAGCTG

[C] GTTCGTAACGT

[D] TTCCGGTATAC

[E] TCCAGGCTAGA

88. A mutation is a change in the sequence of a DNA strand. When studying the genetics of a bacterium, a scientist found its DNA sequence to be AGTTCGCTATCCA. After irradiating the bacterium, the DNA sequence was AGTTCTATCCA. What type of mutation has occurred?

[A] Frame-shift mutation

[B] Nonsense mutation

[C] Deletion mutation

[D] Missense mutation

[E] Reciprocal translocation mutation

89. The DNA sequence of a cell is GCCGTATAGCA. What would be the corresponding strand of mRNA to attach to this strand during transcription?

 [A] CGGCATATCGT

 [B] CGGCAUAUCGU

 [C] AUUACGCGAUC

 [D] TCCTCGCGATA

 [E] GAAUCGACGUA

90. What is the structure used by some prokaryotic cells to propel themselves through their liquid environment?

 [A] Cilia

 [B] Flagella

 [C] Peptidoglycan

 [D] Pilus

 [E] Pseudopodi

Questions 91 – 94 refer to the following illustrations.

A.

B.

C.

D.

91. Where is one most likely to find the structure labeled D?

[A] Inside of the gall bladder

[B] Within the cell's nucleus

[C] Outside the cell membrane.

[D] Within the endoplasmic reticulum

[E] Between the bilayer of lipids

92. Molecule B is produced at the end of which biochemical process?

 [A] ATP synthesis

 [B] Transcription

 [C] Photosynthesis

 [D] Cellular respiration

 [E] Dehydration synthesis

93. What is the function of molecule C?

 [A] Long term energy storage

 [B] Short term energy boost

 [C] Bone construction and destruction

 [D] Transmission of electrical impulses

 [E] Production of hair and fingernails

94. Structure A is a polysaccharide. It cannot be digested by most animals and is often found making up the cell walls of plants. What is it?

 [A] Cellulose

 [B] Dextrose

 [C] Glycogen

 [D] Sucrose

 [E] Starch

95. Many chemical reactions that occur within living things produce energy. They can be represented A + B → AB + energy. What type of reaction is this?

 [A] Anabolic

 [B] Dehydration

 [C] Endothermic

 [D] Exergonic

 [E] Hydrolization

96. No matter what type of molecule is being taken into the cell through the cell membrane, which of these is always needed?

 [A] Membrane invagination

 [B] Phagosomes

 [C] Phagocytosis

 [D] Pinocytosis

 [E] Receptor proteins

97. How does active transport move molecules?

 [A] It moves them from an area of low pH to an area of high pH.

 [B] It moves them in the direction of a higher osmotic potential of the cell.

 [C] It moves them in a direction that is most likely to achieve equilibrium.

 [D] It moves them from an area of high concentration to an area of low concentration.

 [E] It moves them in the direction opposite to the one diffusion moves them.

98. What would happen to a plant cell that loses a large concentration of its water?

 [A] The cell will divide into smaller cells.

 [B] The cell will show a loss of turgor pressure.

 [C] The cell will release its waste products into the cytoplasm.

 [D] The cell will decrease in size equal to the amount of water loss.

 [E] The cell will keep its rigidity due to the cell membrane against the cell wall.

Questions 99 and 100 refer to the following information:

 [A] Prophase I

 [B] Metaphase I

 [C] Metaphase II

 [D] Telophase II

 [E] Interphase

99. During which phase of meiosis are adjacent pieces of homologous pairs going to trade places in order to increase the genetic diversity of the offspring?

100. In oogenesis, when does one parent cell produce one cell with all of the genetic material and three "dead" polar bodies?

101. An organism was found to have DNA that contained 20% cytosine. Which of the following can be concluded about the DNA of this organism?

 [A] It has 30% guanine.

 [B] It has 80% guanine.

 [C] It has 40% adenine and 40% thymine.

 [D] It has 20% thymine and 20% adenine.

 [E] It has 30% thymine and 30% adenine.

102. What can be concluded about the relationship between light intensity and the rate of photosynthesis, as seen in the graph?

 [A] If this plant were given more light, the rate of photosynthesis will continue to increase.

 [B] Photosynthesis is not dependent upon light intensity, but rather the temperature of the air.

 [C] The rate of photosynthesis increases to a point of saturation regardless of how much light it is given.

 [D] If light intensity were decreased, the rate of photosynthesis would continue to rise due to other external factors.

 [E] An increasing light intensity can cause the rate of photosynthesis to fluctuate up to a certain point, and then cause it to drop off.

103. What is the function of aquaporins found the in the cell membrane?

 [A] They protect the inner parts of the cell.

 [B] They allow water to move back and forth.

 [C] They synthesize proteins from amino acids.

 [D] They package materials for export from the cell.

 [E] They eliminate cellular wastes from metabolism.

104. Gregor Mendel was an Austrian monk who studied the inheritance patterns of different characteristics of peas. He discovered that green coloring (G) is dominant to yellow (g) and that round seeds (R) are dominant to wrinkled (r). What would be the projected phenotypic ratio for a cross between two parents that were heterozygous for both traits?

 [A] All peas would be green and round.

 [B] All peas would be yellow and wrinkled.

 [C] Half of the peas would be green/round and half would be yellow/wrinkled.

 [D] There would be 9 green/round, 3 green/wrinkled, 3 yellow/round, 1 yellow/wrinkled.

 [E] There would be 1 green/round, 3 green/wrinkled, 3 yellow/round, 9 yellow/wrinkled.

105. Water is an essential molecule needed by all life on Earth. Which property of water makes it so useful?

[A] It is a polar ionic molecule.

[B] It is a polar covalent molecule.

[C] It is a nonpolar ionic molecule.

[D] It is a nonpolar metallic molecule.

[E] It is a nonpolar covalent molecule.

106. Which element was probably absent from the Earth's early atmosphere?

[A] Hydrogen (H^+)

[B] Oxygen (O_2)

[C] Sulfur (S)

[D] Carbon dioxide (CO_2)

[E] Carbon monoxide (CO)

107. What is the main difference between fermentation and glycolysis?

[A] Fermentation breaks down sugars, while glycolysis breaks down proteins.

[B] Fermentation requires an anaerobic environment, while glycolysis requires an aerobic one.

[C] Fermentation produces six molecules of ATP, while glycolysis produces only two.

[D] Fermentation releases carbon dioxide into the atmosphere, while glycolysis releases oxygen.

[E] Fermentation creates enormous stores of energy, while glycolysis uses large amounts of energy.

108. During cellular respiration, which organelle is responsible for creating a concentration across its membranes?

 [A] The nucleus

 [B] The mitochondria

 [C] The golgi apparatus

 [D] The nucleolus

 [E] The endoplasmic reticulum

109. Which statement correctly identifies the effect of mutations on an organism's DNA?

 [A] They are irreversible.

 [B] They are a source of variation.

 [C] They are only useful when found in germ cells.

 [D] They create selective pressures that cause evolution.

 [E] They benefit the species more than the organism.

110. Which of the following are present in both prokaryotic and eukaryotic cells?

 [A] Cell wall, mitochondria, DNA

 [B] Mitochondria, ribosomes, RNA

 [C] Cell wall, DNA, nuclear membrane

 [D] DNA, RNA, ribosomes, cell membrane

 [E] Cell membrane, nuclear membrane, Golgi apparatus

111. **In mammals, nerve cells have spaces between them, across which signals must be transferred. How do these signals get across this gap?**

 [A] Neural impulses cause the release of chemicals that diffuse across the gap.

 [B] Sodium and potassium rapidly flux back and forth to carry the signals across the gap.

 [C] Electrical currents of varying voltages are emitted from one side of the gap to another.

 [D] The calcium within the axons and dendrites of nerves adjacent to a gap acts as the messenger.

 [E] Esterase serves as the messenger between nerve cells.

112. The following pedigree illustrates the pattern for an autosomal (non-sex-linked) gene that displays complete dominance. This pedigree only shows affected (black) and non-affected individuals.

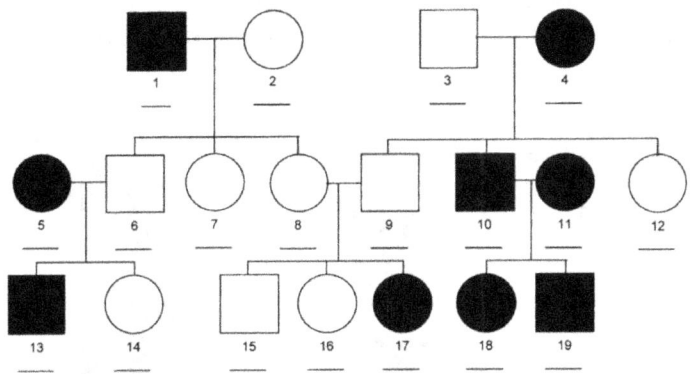

Which members of the above pedigree could be homozygous dominant?

[A] 2 and 3

[B] 2, 9, and 12

[C] 6, 7, and 8

[D] 2, 15, and 16

[E] 8, 9, and 15

113. The fossil record shows that large flying insects arose approximately 250 million years ago. Evidence also suggests that the oxygen concentration in the atmosphere at this time was much higher than it is now, approximately 28%. Scientists investigated the impact of oxygen concentrations on the growth of fruit flies.

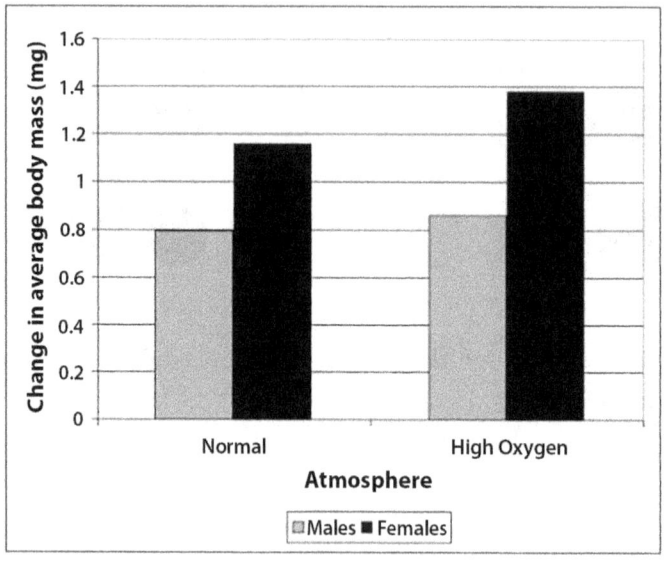

Based on the findings from the study shown in the graph, what would be a reasonable conclusion for the experiment?

[A] The increased sizes of insects were due to higher oxygen concentrations of the primitive atmosphere.

[B] The number of insect species was dependent on the existence of the higher oxygen concentrations.

[C] The metabolism of the large insects was the main reason ancient atmospheres had higher oxygen concentrations.

[D] The higher temperatures and oxygen concentrations contributed to the evolution of large animals on ancient Earth.

[E] Higher temperatures diluted the effects of higher oxygen on the body size of male insects.

114. **The worldwide population of the North Atlantic right whale has fallen so much the species is almost extinct. It also shows very little genetic diversity. Which of these is the most important risk factor the whales face?**

 [A] The habitat of the whales is threatened by climate change and global warming.

 [B] A dwindling food supply makes it more difficult for the right whales to survive and reproduce.

 [C] Mutations are more likely to affect more individuals in populations with low genetic diversity.

 [D] The low genetic diversity interferes with the ability of the population to respond to environmental changes.

 [E] The small population makes it too difficult for whales to find mates.

115. **A scientist wants to find fossils from dinosaurs that lived at the end of the Jurassic period. Where would be a reasonable place to look?**

 [A] Igneous rocks from the late Triassic period (the era just preceding the Jurassic)

 [B] Sedimentary rocks from the late Triassic period (the era just preceding the Jurassic)

 [C] Igneous rocks from the early Cretaceous period (the era just after the Jurassic)

 [D] Sedimentary rocks from the early Cretaceous period (the era just after the Jurassic)

 [E] Metamorphic rocks from the late Triassic period (the era just preceding the Jurassic)

ANSWER KEY

Question Number	Correct Answer	Your Answer	Question Number	Correct Answer	Your Answer	Question Number	Correct Answer	Your Answer
1.	A		40.	E		79.	A	
2.	D		41.	A		80.	C	
3.	B		42.	C		81.	A	
4.	B		43.	B		82.	C	
5.	B		44.	D		83.	E	
6.	C		45.	A		84.	B	
7.	E		46.	C		85.	D	
8.	D		47.	C		86.	A	
9.	B		48.	A		87.	E	
10.	A		49.	D		88.	C	
11.	B		50.	E		89.	B	
12.	C		51.	B		90.	B	
13.	D		52.	B		91.	B	
14.	E		53.	A		92.	C	
15.	A		54.	D		93.	A	
16.	B		55.	D		94.	A	
17.	A		56.	C		95.	D	
18.	D		57.	A		96.	A	
19.	C		58.	E		97.	E	
20.	D		59.	B		98.	B	
21.	D		60.	C		99.	A	
22.	C		61.	D		100.	D	
23.	B		62.	A		101.	E	
24.	D		63.	C		102.	C	
25.	C		64.	B		103.	B	
26.	B		65.	E		104.	D	
27.	E		66.	A		105.	B	
28.	A		67.	D		106.	B	
29.	D		68.	D		107.	B	
30.	D		69.	C		108.	B	
31.	E		70.	E		109.	B	
32.	A		71.	A		110.	D	
33.	D		72.	A		111.	A	
34.	D		73.	D		112.	D	
35.	D		74.	C		113.	A	
36.	B		75.	C		114.	D	
37.	B		76.	C		115.	D	
38.	C		77.	B				
39.	D		78.	E				

Sample Test Two

CLEP Biology Sample Exam 2 Explanations

1. **What is the main function of enzymes?**

 [A] They lower the activation energies of chemical reactions, thereby speeding them up.

 [B] They activate certain hormones within bacterial cells to decrease their rate of binary fission.

 [C] They increase of net gain of ATP produced during cellular respiration in order to improve cell function.

 [D] They convert light energy into chemical energy during photosynthesis so a plant can then produce sugars.

 [E] They remove urea from the excretory system so toxicity levels within the organism do not reach dangerous levels.

 The answer is A.

 Enzymes are specialized proteins that are used by living things to speed up reactions. They accomplish this by lowering the activation energy of the reaction so it can move along faster.

2. **What is the term given to new species colonizing an area over time after a natural disaster?**

 [A] Secondary succession

 [B] Interspecific competition

 [C] Intraspecific competition

 [D] Primary succession

 [E] Tertiary succession

 The answer is D.

 After a natural disaster, such as a volcanic eruption, lichens will be one of the first pioneer species into the area. They grow on rocks. The acids they secrete break down the rocks turning them, and the dead lichens into soil. This soil serves as the basis for small shrubs and plants to colonize the area. Over time, larger plants move in. Finally, tall trees will take over, turning the once-barren area into a climax community.

3. What kind of growth curve is being shown in the graph for a population of bacteria in a Petri dish?

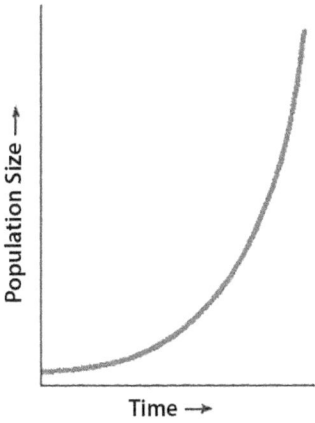

[A] Logistical growth

[B] Exponential growth

[C] Preferential growth

[D] Vertical growth

[E] Redundancy growth

The answer is B.

The population of bacteria is showing exponential growth. This means that one organism divides into two, two into four, four into eight, and so one. The population doubles with each passing interval of time.

4. Over a 20-year span of time, a female whale produced all male offspring. She is pregnant again. What is the probability that this next offspring will be male?

[A] 10%

[B] 50%

[C] 60%

[D] 75%

[E] 100%

The answer is B.

Since whales have two possible genders, there is always a 50/50 chance of male versus female. Having several male offspring in a row does not increase the chances of producing a female calf. It still is all chance.

Questions 5 and 6 refer to the following information:

Achondroplasia is a dominant genetic trait that causes dwarfism (stunted growth). The homozygous condition is lethal. Heterozygotes, however, express the dwarf trait. Rupert has a family history of dwarfism, but does not express it himself. His wife, Lola, has achondroplasia.

5. What are the respective genotypes of Rupert and Lola?

[A] Aa, aa

[B] aa, Aa

[C] Aa, AA

[D] aa, AA

[E] Aa, Aa

The answer is B.

To not show achondroplasia and still be alive, Rupert needs to be homozygous recessive (aa). Lola is showing it, so she has to be heterozygous (Aa).

6. **What is the probability their child will have achondroplasia?**

 [A] 0 %

 [B] 25%

 [C] 50%

 [D] 75%

 [E] 100%

 The answer is C.

 Rupert cannot pass the gene along to his offspring because he does not have one. Lola has a 50% chance of passing along either the dominant or the recessive allele. Therefore, the genotype of their offspring would either be aa (no dwarfism) or Aa (dwarfism).

7. **What type of behavior is a bean plant exhibiting when it bends towards the light?**

 [A] Positive photokinesis

 [B] Negative phototaxis

 [C] Positive phototaxis

 [D] Negative gravitropism

 [E] Positive phototropism

 The answer is E.

 A tropism is the directional movement of an organism. The prefix *photo-* means light. Since the plant is bending in the direction of the light, the response is said to be positive.

8. **A person with type AB blood can receive a transfusion from a person with which blood type(s)?**

 [A] Type A only

 [B] Type A and B

 [C] Type B only

 [D] Type A, B, and O

 [E] Type O only

 The answer is D.

 A person with type AB blood is considered a universal receiver. This means they have all of the possible antigens present within them. They can receive blood from a type A person, a type B person, a type AB person, or a type O person.

Questions 9 and 10 relate to the data table below:

Oxygen Production	
Distance From Light (cm)	Bubbles Produced per Minute
10	39
20	22
30	8
40	5

9. Based upon the finding from this study, what conclusion can be drawn about why primary producers have to live within the uppermost regions of the oceans, ponds, and lakes?

[A] They have to live close to the surface in order to hunt for food.

[B] They have to live close to the surface in order to obtain the light they need for photosynthesis.

[C] They have to live close to the surface in order to produce enough energy to reproduce.

[D] They have to live close to the surface in order to excrete toxic levels of carbon dioxide.

[E] They have to live close to the surface in order to maximize their ability to absorb oxygen from the air.

The answer is B.

Primary producers are highly dependent upon light to make food for themselves. Through photosynthesis, they take in carbon dioxide and water and produce sugars, which are used for energy. A byproduct of this process is oxygen, as represented by the bubbles in the data. The closer the producer is to the light source, the more photosynthesis takes place, and the more bubbles will be produced.

10. **Which is the correct process being supported by the data?**

 [A] $H_2O + CO_2 \xrightarrow{light} sugars + O_2$

 [B] $sugars + O_2 \xrightarrow{light} H_2O + CO_2$

 [C] $H_2O + O_2 \xrightarrow{light} sugars + CO_2$

 [D] $sugars + CO_2 \xrightarrow{light} H_2O + O_2$

 [E] $O_2 + CO_2 \xrightarrow{light} CO_2 + sugars$

 The answer is A.

 The process being supported by the data is photosynthesis. In this biochemical reaction, producers take in water and carbon dioxide and convert them into sugars. Oxygen is a byproduct.

11. **The wing of a bat, the human hand, the fin of a whale, and the front leg of a dog all share a similar bone structure. How do these traits support the theory of evolution?**

 [A] They show diversity between species.

 [B] They show a common ancestor among all species.

 [C] They show how Lamarckism can pass traits between species.

 [D] They show the existence of a unified anatomical theme of all organisms.

 [E] They show how adaptive radiation can lead to the formation of new traits.

 The answer is B.

 These commonalities are called homologous structures. While the adult forms of birds, whales, dogs, and humans all look differently, they are all vertebrates with similar basic anatomy.

12. **Which of the following represents the most likely timeline of the evolution of life on Earth, starting with the oldest?**

 [A] Plants → fungi → animals → humans

 [B] Fungi → bacteria → plants → protists

 [C] Bacteria → protists → plants → animals

 [D] Protists → humans → animals → plants

 [E] Humans → animals → plants → fungi

 The answer is C.

 The first organisms on Earth are believed to have been bacteria. These single-celled organisms specialized, became more complex, and evolved into protists. Protists are either plant-like or animal-like. The plant-like protists migrated onto land and became plants. The animal-like protists evolved into animals.

13. **Which of the following correctly lists the embryological stages of development in animals?**

 [A] Zygote, blastula, gastrula, morula

 [B] Zygote, blastula, morula, gastrula

 [C] Zygote, gastrula, blastula, morula

 [D] Zygote, morula, blastula, gastrula

 [E] Zygote, morula, gastrula, blastula

 The answer is D.

 After fertilization, the zygote starts to divide and become a morula. This happens very quickly and there is no growth between divisions. The morula is a solid ball of cells that then continues to divide into a blastula, which is a hollow ball of cells about one cell thick. The blastula then folds in during a process called gastrulation to become the gastrula with three layers of cells (ectoderm, mesoderm, and endoderm).

14. Monarch butterflies have an orange and black coloration that identifies them well known to predators. Since these butterflies eat milkweed, which is poisonous to other creatures, the predators know to stay away from them. The viceroy butterfly also has an orange and black coloration that is very similar to the monarch's, but the viceroy butterfly does not feed on milkweed, so it is not toxic. Even so, predators still avoid the viceroy. What is the adaptation called that keeps predators from eating the viceroy butterflies?

[A] Mutation

[B] Learning

[C] Reproductive isolation

[D] Pattern formation

[E] Mimicry

The answer is E.

To mimic something means to look a lot like it. The viceroy butterfly looks almost exactly the same as a monarch. Since it only takes one try for predators to realize that monarchs do not taste good, they are likely to remember the color pattern for a long time and avoid eating anything that bears that pattern.

15. If one parent has the genotype AABBCCDDEE and the other parent has the genotype aabbccddee, what are the possible genotypes of their offspring?

[A] All AaBbCcDdEe

[B] AABBccDDEE and aaBBCCDDEE

[C] aaBBccDDee and AAbbCCddEE

[D] AaBBCcDDEE and aABbCCDdEE

[E] AAbbCCddee and AAbbCCddee

The answer is A.

Each parent is homozygous for all of the traits. This means that the first parent can only pass on a dominant allele, and the second parent can only pass on a recessive allele. Therefore, the genotype of all of the offspring has to be heterozygous for all traits.

Number 16–19. Use the following options to answer the questions.

[A] Lipids

[B] Proteins

[C] Carbohydrates

[D] Nucleic Acids

16. Which macromolecule has monomers called amino acids?

The answer is B.

Proteins are made from amino acids. They join together into long chains called polypeptides.

17. Which macromolecule does the body use to store energy for the long term?

The answer is A.

Lipids are used for long-term energy storage. The bonds that hold them together are fairly stable.

18. **Which of these contains instructions for all of the cell's activities?**

 The answer is D.

 Nucleic acids (DNA and RNA) carry the instructions for all living things. DNA makes up chromosomes, which carry the blueprint of life from parent to offspring in sexually reproducing organisms.

19. **Which macromolecule is good for giving a quick burst of energy?**

 The answer is C.

 Carbohydrates (also called sugars) are used by living things for energy. The bonds that hold them together are not as strong as those in lipids, so their energy cannot be stored for as long a time.

20. **Which of the following do NOT exhibit meiotic cell division?**

 [A] Mushroom

 [B] Tree

 [C] Dog

 [D] Bacteria

 [E] Amoeba

 The answer is D.

 Since bacteria do not have a nucleus, they do not undergo mitosis or meiosis. Bacteria only have one circular chromosome, so they are unable to perform crossing over.

21. Which of these refers to the random distribution of maternal and paternal chromosomes into daughter cells?

 [A] Translocation

 [B] Crossing-over

 [C] Non-disjunction

 [D] Independent assortment

 [E] Dominance and recessiveness

 The answer is D.

 The random passing of chromosomes happens through independent assortment. As the chromosomes separate during meiosis, it is unknown which one, the mother's or father's, will be passed into the daughter cell.

22. Which of these organisms is most closely related to the one with the scientific name *Felis concolor*?

 [A] *Canis lupis*

 [B] *Paris bicolor*

 [C] *Felis domesticus*

 [D] *Paris atricapulis*

 [E] *Callinectes sapidus*

 The answer is C.

 The scientific name of the North American mountain lion is *Felis concolor*. Since organisms that are closely related have similar genus names, the most closely related animal to *Felis concolor* would be *Felis domesticus*, the house cat.

23. **What characteristic is the distinguishing feature of birds?**

 [A] Birds are able to lay eggs.

 [B] Birds have bodies covered with feathers.

 [C] Birds migrate to warmer climates during the winter.

 [D] Birds build complex nests to protect their young.

 [E] Birds eat many different types of foods for energy.

 The answer is B.

 All birds are covered with feathers. No other animals have feathers, so this makes them the distinguishing characteristic of all birds. Reptiles build nests and also lay eggs. Many species of fish and mammals migrate to warmer climates in the winter. Many species are opportunists when it comes to finding food, eating whatever they can find.

24. **In an experiment, lab mice had their tails snipped soon after birth for 15 generations. When the mice were reproduced for a 16th time, all of the offspring had tails. What conclusion can be drawn from these results?**

 [A] The mice were showing natural selection.

 [B] The mice were exhibiting the principles of Lamarckism.

 [C] The mice showed little or no similarities to their parents.

 [D] The mice showed that mutations in gametes could affect offspring traits.

 [E] The mice showed that offspring inherit characteristics acquired by their parents.

 The answer is D.

 Since all generations of mice had tails, Lamarckism was not at work here. This idea says that traits acquired by parents can be passed onto their offspring. If this were the case, then none of the mice would have tails. Since only the chromosomes of the gametes are inherited by the offspring, only changes in those DNA would produce changes in phenotypes.

Questions 25 – 29 relate to the following organs found within mammals.

[A] Stomach

[B] Small intestine

[C] Liver

[D] Pancreas

[E] Large intestine

25. **Which organ creates the enzyme that breaks down fats?**

 The answer is C.

 The liver creates an enzyme called bile that is used to emulsify fats when they enter the small intestine.

26. **Which organ is the site of most chemical digestion?**

 The answer is B.

 The small intestine is where most chemical digestion takes place. It has small, finger-like structures lining its walls that increase its surface area and that help with absorption. Also, the gall bladder and pancreas both release their enzymes into this organ to assist with the breakdown of food.

27. **Which organ is responsible for the absorption of water from the waste products of digestion?**

 The answer is E.

 The large intestine is where water and some nutrients are reabsorbed. Once the water has been removed, compaction occurs to prepare the waste products for elimination from the body.

28. **Which organ breaks apart and grinds up food that enters it from the mouth?**

 The answer is A.

 The stomach is a muscular organ that is responsible for grinding up food. It also secretes a few enzymes that aid with the chemical digestion of proteins and carbohydrates.

29. **Which organ secretes the enzyme that is mainly responsible for the digestion of proteins?**

 The answer is D.

 The pancreas is an organ that works with both the digestive and endocrine systems. In the digestive system, it produces trypsin, an enzyme that helps to break down proteins in the small intestine. For the endocrine system, the pancreas produces insulin, a hormone needed to maintain blood sugar levels.

30. **Which of the following is true regarding the flow of energy and nutrients through an ecosystem?**

 [A] Energy and nutrients are both recycled.

 [B] Both energy and nutrients accumulate in the highest trophic levels.

 [C] Only 10 % of the available energy is recycled due to loss from metabolism.

 [D] Nutrients get recycled, but most of the energy is lost from every trophic level.

 [E] The amount of nutrients available for each tropic level is depends on the available energy.

 The answer is D.

 While it not possible for energy to be removed from a system, it does get converted into other forms that make it unavailable for organisms to use. As energy moves up the trophic pyramid, it gets used for an organism's metabolism. In fact, 90% of the energy is used, so only 10% can be moved up. Nutrients, on the other hand, can be recycled. Carbon, nitrogen, phosphorus, and so on, all get moved around an ecosystem.

31. Which of the following BEST explains why there are usually fewer than five trophic levels to most food chains?

[A] Many primary consumers feed at more than one trophic level.

[B] The carrying capacity of the environment would be exceeded with more than five levels.

[C] Ecosystems with more than five levels contain too much biomass.

[D] The increased demand on the tertiary consumers would cause them to face extinction.

[E] Each trophic level only obtains a small fraction of the energy from the trophic level below it.

The answer is E.

In a food chain, there is a principle called the 10% rule. This means that only 10% of an organism's energy is available to the next trophic level. Most of the energy is used by the organism itself to maintain its metabolism. Since there is so little energy available for the higher trophic levels, very few food chains can support more than five levels.

Questions 32–34 refer to the following information:

Humans can have either attached earlobes (f) or free earlobes (F). Assume that two parents are heterozygous for free earlobes.

32. What are the genotypes of the parents?

[A] Ff x Ff

[B] Ff x ff

[C] ff x Ff

[D] FF x ff

[E] ff x FF

The answer is A.

Both parents have a dominant and a recessive allele. This is what makes them heterozygous.

33. **What percentage of the parent's sex cells carried the allele for free earlobes?**

 [A] 0%

 [B] 25%

 [C] 50%

 [D] 75%

 [E] 100%

 The answer is D.

 Since both parents have a dominant and a recessive allele for earlobes, when meiosis occurs, each parent can pass on either one. This makes the probability of inheriting the dominant 50% allele and the recessive allele 50%.

34. **What are the possible phenotypes of the offspring?**

 [A] All offspring will have free earlobes.

 [B] All offspring will have attached earlobes.

 [C] 25% of the offspring will have free earlobes.

 [D] 75% of the offspring will have free earlobes.

 [E] 100% of the offspring will have free earlobes.

 The answer is D.

 After doing the Punnett square, you can see that there is a 25% chance of a FF genotype, a 50% chance of an Ff genotype, and a 25% chance of an ff genotype. Since, to have free earlobes, only one dominant gene s needed, those boxes having an F are free lobes. Therefore, there is a 75% chance of free earlobes in the offspring.

35. Which of the following is an example of a polysaccharide?

 [A] Saturated fats

 [B] Cholesterol

 [C] Glucose

 [D] Starch

 [E] Lysine

The answer is D.

Starch is the carbohydrate used by plants to store energy. It is a huge molecule, thereby making it a polysaccharide.

36. What would most likely happen to a plant cell that was placed into an isotonic solution?

 [A] It would become turgid.

 [B] It would become flaccid.

 [C] It would swell and lyse.

 [D] It would elongate.

 [E] It would undergo apoptosis.

The answer is B.

Plant cells have a cell wall, which prevents lysing (rupture of the cell membrane). Isotonic environments do not allow for swelling or turgidity, so plant cells become flaccid.

37. In snapdragons, red flowers are dominant to white. A researcher did the following cross and got the reported results.

red snapdragon x white snapdragon = pink snapdragon

pink snapdragon x pink snapdragon = red, white, and pink snapdragon

What is the possible explanation for these results?

[A] Snapdragons show a condition called codominance, where both alleles show up at the same time.

[B] Snapdragons demonstrate incomplete dominance, which causes phenotypes to mix in heterozygotes.

[C] Snapdragons can express different colors, which can be used as indicators of environmental conditions.

[D] Snapdragons are prone to mutation, so when pink flowers were crossed the different colors were expressed.

[E] Snapdragons are all heterozygous, which leads to the expression of the dominant and recessive traits when crossed.

The answer is B.

Incomplete dominance means that phenotypes get mixed in heterozygote snapdragon plants. In this case, the red x white made pink. When pink flowers (heterozygotes) are crossed, the dominant and recessive phenotypes are expressed along with the incomplete dominant one.

38. **After competing in a long race, a runner continues to breathe hard even after crossing the finish line. What purpose does this serve?**

 [A] It causes the brain to return to a normal resting state.

 [B] It helps the runner to produce more glucose for energy.

 [C] It repays the oxygen debt created from muscle exertion.

 [D] It distributes energy around the body to the runner's organs.

 [E] It drops the runner's blood pressure so the heart rate returns to normal

 The answer is C.

 When muscles are used for a long period of time, the stored oxygen within them gets used up. After the use stops, the body needs to repay the deficit of oxygen, so the brain and heart keep the body breathing hard for a period of time afterwards to resupply the muscles with oxygen.

39. **Which of the following predicts what would happen if a participant in a study were given air to breathe that had a higher than usual carbon dioxide concentration?**

 [A] They would have increased respiration and heart rate

 [B] They would have decreased respiration and blood pH.

 [C] They would have decreased respiration and heart rate.

 [D] They would have increased respiration and decreased blood pH.

 [E] They would have decreased respiration and increased blood pH.

 The answer is D.

 Carbon dioxide is acidic. Breathing in a high concentration of this gas would cause the blood pH to decrease, becoming more acidic. Also, since the brain would be detecting more carbon dioxide in the blood, it would try to stabilize the pH by bringing in more oxygen, thus increasing the person's breathing rate.

40. In terms of evolution, which type of organism would be considered the most successful?

 [A] An organism that has the largest territory.

 [B] An organism that eats the greatest variety of food.

 [C] An organism that can reproduce the most often.

 [D] An organism that has the largest amount of biomass.

 [E] An organism that leaves behind the greatest number of offspring.

 The answer is E.

 If an organism has been very successful, it has produced many children and passed on its genes many times. These offspring have also reproduced and passed on their genes, which are mostly the genes from their parents.

41. What is the structure called that plants use to protect their leaves from water loss?

 [A] Cuticle

 [B] Dermis

 [C] Parenchyma

 [D] Phloem

 [E] Xylem

 The answer is A.

 Land plants produce a waxy cuticle that covers their leaves. This covering is waterproof, so it prevents the loss of water through the tissue layers of the leaves.

42. During the fall, many deciduous trees change color. For example, sugar maples of New England tend to go from green to bright oranges, yellows, and reds. What is the most likely explanation for this?

[A] The absorption spectrum of chlorophyll changes during the fall to include green wavelengths.

[B] The trees increase their production of these pigments to adapt to the different amounts of light.

[C] There is a reduction in the production of green chlorophyll, so the masked pigments become visible.

[D] The light available at this time of year lacks the blue and green wavelengths found in the summer light.

[E] There are more orange, yellow, and red wavelengths found in the light during the fall than in the spring and summer.

The answer is C.

Chlorophyll is the pigment that makes leaves green. During the fall there is less light available, so chlorophyll production decreases. As a result, the green color of leaves disappears, and the masked pigments become visible.

Questions 43–46 refer to the following choices:

[A] Mutualism

[B] Commensalism

[C] Parasitism

[D] Predator/prey

[E] Competition

43. **What happens when two species live together and one of them benefits but the other is unharmed?**

 The answer is B.

 Commensalism happens when two species live together but only one of them benefits without hurting the other. For example, barnacles living on the rostrum of a humpback whale exhibit this type of relationship. The barnacles get a free ride to wherever the whale goes and leftover food scraps from it. The whale most likely does not even know the barnacles are there.

44. **What type of relationship exists when one species hunts and eats another?**

 The answer is D.

 A predator is an organism that hunts and eats other organisms, for instance, a lion hunting a gazelle on the African plains. This prey is killed to benefit the predator.

45. **What type of relationship do a clownfish and a sea anemone exhibit?**

 The answer is A.

 The clownfish and the sea anemone are a classic example of a mutualistic relationship. The clownfish lives within the tentacles of the anemone. This offers it protection from predators. In return, the sea anemone gets leftover food particles from the clownfish. Both animals benefit.

46. **What type of relationship exists between a tick and a deer?**

 The answer is C.

 The tick/deer relationship is an example of parasitism. In this case, one organism benefits (the tick) at the expense of the other (the deer). The tick is sucking the blood out of the deer. While doing so, it could potentially transmit disease. In other cases of parasitism, the host may die. If this happens, the parasite needs to move to another host for its survival.

47. Protein synthesis always results in the production of which of these?

 [A] Ammonia

 [B] ATP

 [C] Water

 [D] Carbon dioxide

 [E] Oxygen

 The answer is C.

 Protein synthesis is the result of a dehydration reaction. Water is a byproduct.

48. What is the mass of chewed food and saliva called that enters the esophagus?

 [A] Bolus

 [B] Duodenum

 [C] Ilium

 [D] Retina

 [E] Trypsin

 The answer is A.

 After the food is chewed and mixed with saliva, it is called a bolus. This mass gets positioned by the tongue at the back of the mouth so it can then pass into the esophagus.

49. **Which of the following is correctly matched with its structure of excretion?**

 [A] Fish – nephridia

 [B] Honeybee – flame bulbs

 [C] Planaria – kidneys

 [D] Grasshopper – malphigian tubules

 [E] Humans – spleen

 The answer is D.

 Fish have kidneys to produce urine. Planarians use flame bulbs to excrete wastes. Honeybees, grasshoppers, and other insects use Malphigian tubules. Humans use their kidneys for excretion.

The phylogenetic tree below traces the evolution of plants. Questions 50 – 54 relate to the phylogenetic tree found below.

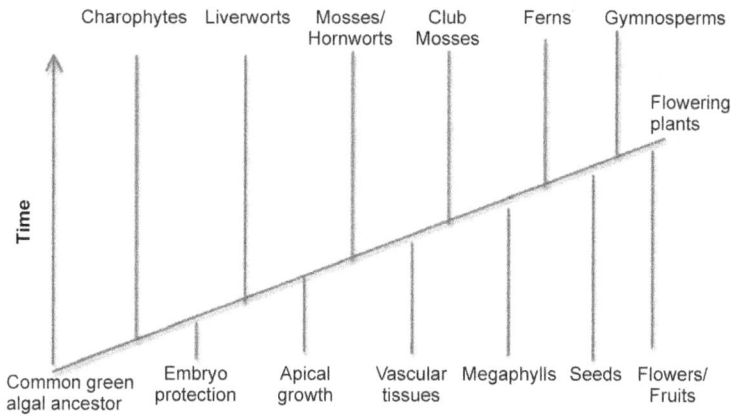

50. Which trait was the earliest to appear in the evolution of plants?

 [A] Seeds

 [B] Flowers

 [C] Apical growth

 [D] Vascular tissues

 [E] Embryo protection

The answer is E.

Plants developed a mechanism to protect their embryos early in their evolution. This showed to be successful, especially when they migrated onto land.

51. **Which two types of plants are most closely related to each other?**

 [A] Hornworts and ferns

 [B] Flowering plants and gymnosperms

 [C] Charophytes and club mosses

 [D] Liverworts and flowering plants

 [E] Mosses and charophytes

 The answer is B.

 The flowering plants and gymnosperms are closest together on the phylogenetic tree. This indicates that their evolution is more closely related than that of other types of plants.

52. **Which plants never evolved the ability to have apical growth?**

 [A] Ferns

 [B] Charophytes

 [C] Hornworts

 [D] Gymnosperms

 [E] Flowering plants

 The answer is B.

 The charophytes that evolved soon after plants moved onto land are the earliest land plants in their evolutionary history. They never developed apical growth and are most closely related to the ancient green algal ancestor.

53. **Which structure has led to the success of the most recent land plants?**

 [A] Fruits

 [B] Seeds

 [C] Megaphylls

 [D] Vascular tissue

 [E] Embryo protection

 The answer is A.

 The most recently evolved plants are the flowering plants. These organisms have developed a structure called a fruit that protects and nourishes the seeds. Fruits are also used to disperse the seeds to new locations, which helps pass on the plant's genes.

54. **What conclusion can be drawn from the information presented in this phylogenetic tree?**

 [A] The most ancient form of land plants is the moss.

 [B] Vascular tissues were essential to the success of all land plants.

 [C] Ferns are the most closely related land plants to the ancient green algal ancestor.

 [D] Liverworts have adapted and survived without the development of seeds or vascular tissue.

 [E] The progression of time has caused some land plants to become less complex than their ancestors.

 The answer is D.

 The phylogenetic tree shows that liverworts evolved well before the development of vascular tissues. Since liverworts are still in existence, that indicates that even without these specialized tissues, they are well adapted for their particular environment.

55. Which of these animals would be considered endothermic?

[A] Bacterium

[B] Crocodile

[C] Grasshopper

[D] Hummingbird

[E] Rattlesnake

The answer is D.

Birds are endothermic. This means they maintain a constant internal body temperature and are not at the mercy of the environment for their metabolism. Insects, like the grasshopper, and reptiles are considered ectothermic. This means they need to warm themselves in the Sun when they are cold and move to a cooler location when they are too warm. Bacteria do not have any type of body temperature.

Questions 56–59 refer to the food web in the diagram

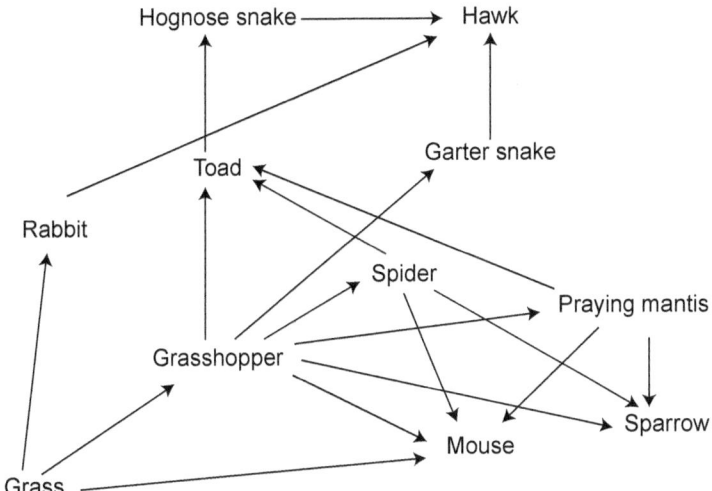

56. Which organisms in this food web would be considered primary consumers?

 [A] Hognose snake, hawk, sparrow

 [B] Spider, rabbit, praying mantis

 [C] Rabbit, grasshopper, mouse

 [D] Grasshopper, toad, hark

 [E] Garter snake, praying mantis, mouse

 The answer is C.

 Primary consumers are those organisms that eat the autotrophs. In this case, the grass is the autotroph, so the rabbit, grasshopper, and mouse are all primary consumers.

57. **In which sequence of events is the mouse considered a tertiary consumer?**

 [A] Grass → grasshopper → spider → mouse

 [B] Grass → rabbit → hawk → mouse

 [C] Grass → mouse → hognose snake → hawk

 [D] Grass → sparrow → praying mantis → garter snake

 [E] Grass → toad → hawk → grasshopper

 The answer is A.

 The mouse occupies the tertiary consumer role in the food chain that starts with the grass and then goes to the grasshopper and then the spider. The mouse then eats the spider.

58. **In this food web, what would be a likely impact if the grasshopper population were to decrease?**

 [A] The number of rabbits would increase.

 [B] The hawk population would decrease as well.

 [C] The hognose snake population would increase.

 [D] The mouse population would show a steady decline.

 [E] The number of garter snakes would drastically decrease.

 The answer is E.

 In this food web, the only things the garter snakes eat are the grasshoppers. If the grasshoppers were to decrease in numbers, the garter snakes would have nothing to eat. Therefore, their population size would dramatically go down.

59. **Why is the toad considered an opportunistic feeder in this food web?**

 [A] The toad gets eaten by several different organisms.

 [B] The toad is part of three different food chains.

 [C] The toad's main predator is the hognose snake.

 [D] The toad's main source of food is the hawk's leftovers.

 [E] The toad can survive under many different environmental conditions.

 The answer is B.

 In this food web, the toad is part of three different food chains. This means that if something happens to one of its food sources, it still has other opportunities to eat.

60. **Which of the following can feed at more than one level of the trophic pyramid?**

 [A] Primary producer

 [B] Secondary consumer

 [C] Omnivore

 [D] Carnivore

 [E] Herbivore

 The answer is C.

 Omnivores are those organisms that eat both plants and animals. This makes them both primary and secondary consumers.

61. Which of the following would be considered a density-independent factor that can cause a change within a population?

 [A] Predation

 [B] Mate selection

 [C] Competition

 [D] Natural disasters

 [E] Niche selection

 The answer is D.

 No matter how large or small a population may be, a natural disaster, like a hurricane or a volcanic eruption, can change its size. Predation, mate selection, competition, and niche selection all depend on how large the population is at the time.

62. Tapeworms live within the intestines of large mammals where they absorb the nutrients intended for the mammal and use them for their own growth and reproduction. Since the large mammal is not getting any nutrition, it starts to lose weight and wither away. What kind of relationship exists between the tapeworm and the mammal?

 [A] Parasitism

 [B] Mutualism

 [C] Commensalism

 [D] Predator/prey

 [E] Ammensalism

 The answer is A.

 The tapeworm is considered a parasite on the large mammal. It benefits from the nutrients it is absorbing but is also harming the mammal, causing it to starve.

Questions 63 – 67 pertain to the different types of biomes presented below.

[A] Savanna

[B] Desert

[C] Tropical rainforest

[D] Temperate forest

[E] Open ocean

63. Which biome has the richest species diversity?

 The answer is C.

 The topical rainforests are known for their species diversity. There are more individual species living in these areas than in several of the other biomes combined, perhaps, it is believed, because of the ideal climate and abundance of resources.

64. Which biome receives the least amount of rainfall every year?

 The answer is B.

 The lack of precipitation in the desert is its distinguishing feature. Most tropical deserts receive less than 20 cm of rain every year.

65. Where would one go to see how light penetration could affect the rates of photosynthesis?

 The answer is E.

 The open ocean can reach great depths, but light does not penetrate that far down into the water. Since photosynthesis can only occur in the upper levels of the water column, all marine producers need to live in areas where light is available.

66. Which biome has low shrubs, tall grasses, and herds of animals?

 The answer is A.

 The savanna is known for its wide-open areas scattered with occasional low shrubs and grasses. Many herding animals find this environment favorable because of the plentiful food sources.

67. **Where would one go to have a mild summer, a cold winter, and see the leaves fall off the trees?**

 The answer is D.

 Temperate forests are located in the northern hemisphere, where they tend to have seasons of cold and warm. Many of the trees in these areas are called deciduous, which means they drop their leaves in the fall.

68. The fossil record can be used to estimate the total number of taxonomic families that have existed over time. These estimates are shown in the graph. The letters A-D represent periods of mass extinctions. What can be concluded from the data?

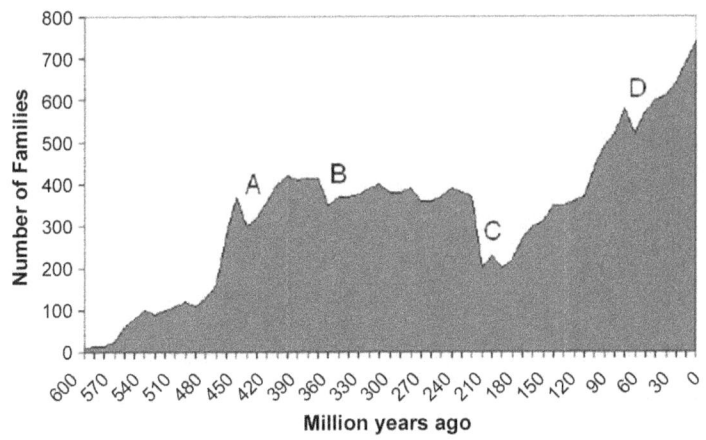

[A] Changes in global climate can result in a mass extinction.

[B] The total amount of biodiversity has difficulty recovering after a mass extinction.

[C] Mass extinctions generally eliminate all life on the planet, so ecosystems need to start over.

[D] Mass extinctions are often followed by periods of speciation and increased biodiversity.

[E] There is incomplete evidence from the fossil record to associate mass extinctions with the biodiversity.

The answer is D.

The graph shows that after a period of mass extinction, there is a period of growth. All four of the major events show this pattern, albeit in differing intensities. The smallest period of speciation was after extinction B and the largest was after extinction C.

CLEP Biology Sample Exam 2 Explanations 347

69. While examining a population of fish in a river ecosystem, you notice that the density of healthy adults has increased to a level much higher than your previous record. What can be concluded from this observation?

 [A] There was a reduction in the death rate.

 [B] There was a reduction in the birth rate.

 [C] There was an increase in the rate of immigration.

 [D] There was an increase in the rate of emigration.

 [E] There was a reduction in the rate of emigration.

 The answer is C.

 Since there are more adult fish in this river than the last time they were observed, it seems logical that the number of fish coming into the area had increased. An increase in the birth rate would be evident if there were more juveniles present.

70. Which of these would be considered a niche of a beetle living in the tropical rainforest?

 [A] The rainforest floor

 [B] Living in the canopy

 [C] The presence of predators

 [D] All of the leaves on a tree

 [E] A particular tree in the forest

 The answer is E.

 An organism's niche is the particular place where it lives and operates. It includes all of the biological and physical conditions it needs in order to survive.

71. A volcanic eruption has just ended, wiping out all of the trees, grasses, and shrubs on a new island. All of the animals are gone as well. What are the first organisms called that will repopulate this area over time?

[A] Pioneer species

[B] Climax species

[C] Primary producers

[D] Keystone species

[E] Primary consumers

The answer is A.

Pioneer species are the first organisms to colonize an area after a disturbance. In this case, primary succession will occur, so the pioneer species will most likely be lichens growing on the rocks left by the explosion.

Use the graph of the Age Structure of the World Population to answer questions 72 and 73.

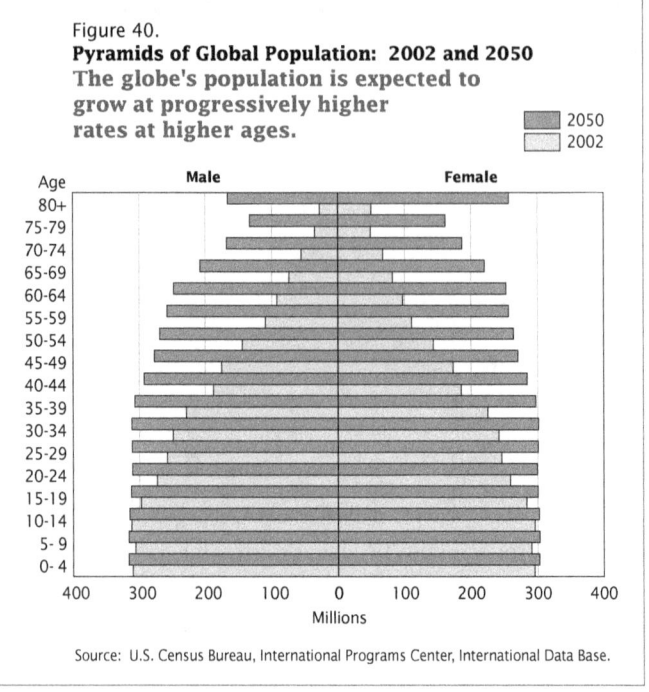

72. Which of these is a trend in the human population between 2002 and 2050?

[A] People are expected to live longer in the future.

[B] There will be consistently more females than males.

[C] Middle age will be found between 55 -59 in 2050.

[D] Children will continue having difficulty living past age 15.

[E] Far more people were alive in 2002 than will be in 2050.

The answer is A.

The population graph shows that from 2002 to 2050 there will not only be an increase in the world's population, but there will also be a shift in the age structure. There will be a much wider distribution of ages. People are expected to live well past 70 years of age.

73. Due to the predicted age structure of the human population, which of the following is likely to eventually happen to the plant's demographics?

 [A] The number of children below 19 will show a decrease.

 [B] There will be an increase in the number of people over 70 years old.

 [C] There will be a decrease in the number of people between 30 and 50 years of age.

 [D] The overall population of the world will decrease because many will be past reproductive age.

 [E] The populations of some countries will continue to increase at dramatic rates due to lack of preventative measures.

 The answer is D.
 According to the graph, in 2050 there will be a large number of people in the middle age to senior citizen age range. Since these people are most likely past the age of their reproductive capabilities, they are unable to have more children. As a result, the overall world population will decrease.

74. Which of these would be considered a renewable resource?

 [A] Coal mining

 [B] Oil production

 [C] Solar energy panels

 [D] Natural gas extraction

 [E] Uranium in radioactive decay

 The answer is C.
 Solar energy is a renewable resource. No matter how much of it gets used, there will always be more. Oil, natural gas, coal, and uranium are resources that have a limited stores. Once they are gone, there is no more.

75. The zebra mussel (*Dreissena polymorpha*) is a small bivalve that has been known to cause problems for boaters in the Great Lakes by attaching to their propellers and clogging up irrigation pipes, preventing water flow. These animals have few natural predators and are not native to the area. What is the term that describes this type of non-native organism?

[A] Biodiversity

[B] Predator

[C] Invasive species

[D] Parasitic species

[E] Polygenic species

The answer is C.

The zebra mussel is an invasive species that arrived in the Great Lakes region most likely in ballast water from a cargo ship. Since it has no natural predators, it is able to survive and reproduce to the point where it has become disruptive.

76. Which of these is believed to be the greatest cause of the greenhouse effect?

[A] The melting of the polar ice caps

[B] The recycling of paper and plastics

[C] The release of carbon dioxide from factories and cars

[D] The rise in ocean levels

[E] The decrease in biodiversity in the tropical rainforests

The answer is C.

Automobiles and factories are the prime contributors to the gases that humans have added to the atmosphere. As the Sun's rays penetrate the earth's atmosphere, they are unable to escape due to this layer of gases. As a result, the heat from these rays stays on Earth and increases the temperature. This increase leads to a melting of the polar ice caps, rising sea levels, and decreases in biodiversity as species are unable to adapt fast enough to changing conditions.

77. All ecosystems are made up of biotic and abiotic factors. Which statement reflects just the biotic aspects?

[A] Salmon can live in both fresh and salt water.

[B] Pandas can only live in areas where bamboo grasses grow.

[C] Whales have low metabolic rates compared to other mammals

[D] The seeds of some pine trees need fire in order to germinate.

[E] Brown algae prefer grow in areas where the water is very cold.

The answer is B.

While all other answers have a biotic and an abiotic factor, only answer B deals with biotic factors. The pandas only eat bamboo grass, so they depend on it for their survival.

78. A farmer's corn crops are being plagued by a new pest species. These insects eat the leaves and all the seeds on the cobs. She wants to decrease the population of these pests in the most ecologically friendly way possible. What should the farmer do?

[A] The farmer should add pesticides to the water used on the crops.

[B] The farmer should introduce a pathogen into the soil to kill the pests.

[C] The farmer should introduce a natural predator of the pest into the field.

[D] The farmer should add additional competitors for the corn into the system.

[E] The farmer should decrease the carrying capacity of the insect pest's habitat.

The answer is E.

By decreasing the carrying capacity of the habitat, the number of pests will gradually decrease. This can be accomplished by removing the resources on which the pest depends.

Use the information in the graph to answer questions 79 and 80.

In a classic experiment to show competition, two species of paramecium are cultured separately in different containers and then again together in the same container. The population sizes of each species are recoded in the graphs below.

79. What conclusion can be made from examining the population graphs of these two species?

[A] *Paramecium auralia* outcompetes *P. caudatum* when grown together.

[B] *Paramecium caudatum* is a better competitor than *Paramecium auralia*.

[C] Both species are able to survive equally well when grown in the same culture.

[D] If cultured together, the species of Paramecium will interbreed to create a more suitable hybrid.

[E] The resources needed for both species to live successfully are in high supply when cultured together.

The answer is A.

Considering the graphs, the population size of *P. auralia* is much larger than *P. caudatum* when the species are cultured together. This is because *P. auralia* is a better competitor, using more of the resources. It actually forces the localized extinction of the other species here.

80. **What ecological principle is at hand when both species are cultured in the same container?**

 [A] Intraspecific competition

 [B] Predator/prey relationship

 [C] Competitive exclusion

 [D] Mutualistic relationship

 [E] Convergent evolution

 The answer is C.

 When both species of Paramecium are grown together, *P. auralia* causes the demise of the *P. caudatum*. *P. auralia* is a better competitor, being larger in size and better able to get the available resources.

Questions 81–85 refer to the diagram below.

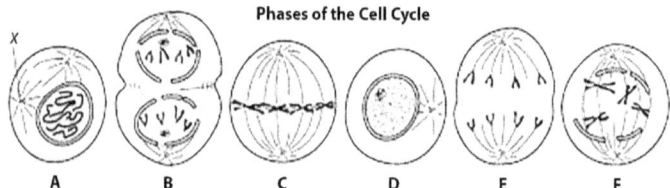

81. **Which represents the cell cycle in the proper order?**

 [A] D, A, F, C, E, B

 [B] A, B, C, D, E, F

 [C] F, E, D, C, B, A

 [D] C, E, B, F, A, D

 [E] B, F, E, A, D, C

 The answer is A.

 The cell cycle goes interphase, prophase, premetaphase, metaphase, anaphase, telophase.

82. **Diagram C shows the chromosomes lined up along the equator. During which phase of the cell cycle does this happen?**

 [A] Anaphase

 [B] Cytokinesis

 [C] Metaphase

 [D] Prophase

 [E] Telophase

 The answer is C.

 During metaphase, the chromosomes line up along the equator of the cell. The spindle fibers will then pull them apart into individual chromatids during anaphase.

83. **What is happening during the phase of the cycle represented by letter D?**

 [A] This is cytokinesis where the cell membrane is being split in half.

 [B] This is prophase where the nuclear envelope is being dissolved.

 [C] This is metaphase where the chromosomes line up along the equator of the cell.

 [D] This is anaphase where the individual chromatids get pulled towards opposite poles.

 [E] This is interphase where the genetic material and all the organelles are being duplicated.

 The answer is E.

 Interphase is the longest part of the cell cycle. Here, all of the cell's genetic material gets copied so that it can be passed to the daughter cells when the cell divides. Since the daughter cells are identical to the parent cell, all of the organelles need to be copied as well.

84. **What is the function of the process shown in the diagram?**

 [A] To breakdown worn out cell parts

 [B] To enable the organism to grow in size

 [C] To encourage development of new traits

 [D] To destroy foreign invaders within the cell

 [E] To create new genetic combinations that cause evolution

 The answer is B.

 The process shown in the diagram is called mitosis. This process, otherwise known as cell division, allows an organism to grow in size. One cell divides into two. Two cells will divide into four. Four divide into eight, and so on. Mitosis also allows an organism to repair damaged cells.

85. **What happens after the cell completes the phase of the diagram shown in B?**

 [A] The cell will sit and rest for a while.

 [B] The cell will jump back into metaphase.

 [C] The cell will divide three more times.

 [D] The cell will enter back into interphase.

 [E] The cell will replace the missing chromatids.

 The answer is D.

 After the cell finishes mitosis by dividing into two daughter cells during telophase, it enters back into interphase. Here, the entire process starts all over again. The cell will grow, duplicate its genetic material, and then grow a little more.

86. **What does mitosis in a plant cell have to deal with that mitosis in an animal cell does not?**

 [A] A cell wall

 [B] A nucleus

 [C] A nucleolus

 [D] A cell membrane

 [E] A golgi apparatus

 The answer is A.

 Plant cells have a cell wall that interferes with mitosis. Since this structure is used for support, it cannot as easily divided as the cell membrane. Instead, plant cells do their division within the cell wall and then form a structure called a cell plate to separate each new cell.

87. **What would the complementary strand be to a piece of DNA with the nucleotide sequence AGGTCCGATCA?**

 [A] AGGTCCGTCA

 [B] GAACTTAGCTG

 [C] GTTCGTAACGT

 [D] TTCCGGTATAC

 [E] TCCAGGCTAGA

 The answer is E.

 According to Chargaff's rule, A goes with T and C goes with G. The complementary strand of the DNA molecule is that matching pair on the other side.

88. A mutation is a change in the sequence of a DNA strand. When studying the genetics of a bacterium, a scientist found its DNA sequence to be AGTTCGCTATCCA. After irradiating the bacterium, the DNA sequence was AGTTCTATCCA. What type of mutation has occurred?

[A] Frame-shift mutation

[B] Nonsense mutation

[C] Deletion mutation

[D] Missense mutation

[E] Reciprocal translocation mutation

The answer is C.

The new sequence of DNA is missing two of the nitrogen bases that were present in the original strand. This means the radiation caused them to be deleted somehow.

89. The DNA sequence of a cell is GCCGTATAGCA. What would be the corresponding strand of mRNA to attach to this strand during transcription?

[A] CGGCATATCGT

[B] CGGCAUAUCGU

[C] AUUACGCGAUC

[D] TCCTCGCGATA

[E] GAAUCGACGUA

The answer is B.

During transcription, the RNA strand matches up with the DNA strand to form a complement. However, in RNA there is no T. It is replaced with U. Therefore, every time A is present in DNA the RNA matches up a U.

90. **What is the structure used by some prokaryotic cells to propel themselves through their liquid environment?**

 [A] Cilia

 [B] Flagella

 [C] Peptidoglycan

 [D] Pilus

 [E] Pseudopodi

 The answer is B.

 The flagellum is a hair-like appendage attached to some prokaryotic cells. It whips back and forth, moving the organism forward.

Questions 91 – 94 refer to the following illustrations.

A.

B.

C.

D.

91. Where is one most likely to find the structure labeled D?

[A] Inside of the gall bladder

[B] Within the cell's nucleus

[C] Outside the cell membrane.

[D] Within the endoplasmic reticulum

[E] Between the bilayer of lipids

The answer is B.

The molecule labeled D is DNA recognizable by its double-helix structure. It is found within a cell's nucleus.

92. **Molecule B is produced at the end of which biochemical process?**

 [A] ATP synthesis

 [B] Transcription

 [C] Photosynthesis

 [D] Cellular respiration

 [E] Dehydration synthesis

 The answer is C.

 Molecule B is glucose. It is produced by autotrophs at the end of the Calvin Cycle in photosynthesis and is broken down by these organisms for energy.

93. **What is the function of molecule C?**

 [A] Long term energy storage

 [B] Short term energy boost

 [C] Bone construction and destruction

 [D] Transmission of electrical impulses

 [E] Production of hair and fingernails

 The answer is A.

 This molecule is a fatty acid. It is part of a lipid, which is used by the body to store energy for the long term. Proteins are used to build hair, fingernails, and bone. Short-term energy boosts are provided by carbohydrates because they are much easier than lipids to break down. .

94. Structure A is a polysaccharide. It cannot be digested by most animals and is often found making up the cell walls of plants. What is it?

 [A] Cellulose

 [B] Dextrose

 [C] Glycogen

 [D] Sucrose

 [E] Starch

 The answer is A.

 This huge molecule is cellulose, made from many repeating sugar molecules. It is found in the cell walls of plants and offers them support to stand upright toward the Sun. Dextrose and sucrose are much smaller sugars. Glycogen is found in animal cells, and starch is another polysaccharide that is found in the roots of plants. This is how they store energy for the long term.

95. Many chemical reactions that occur within living things produce energy. They can be represented A + B → AB + energy. What type of reaction is this?

 [A] Anabolic

 [B] Dehydration

 [C] Endothermic

 [D] Exergonic

 [E] Hydrolization

 The answer is D.

 In exergonic reactions heat is produced and will appear as one of the products in the equation. Hydrolytic reactions break things down by adding water into the system. Dehydration reactions have water as a product. Anabolic reactions take in heat.

96. No matter what type of molecule is being taken into the cell through the cell membrane, which of these is always needed?

 [A] Membrane invagination

 [B] Phagosomes

 [C] Phagocytosis

 [D] Pinocytosis

 [E] Receptor proteins

 The answer is A.

 Water, food particles, and proteins are taken into the cell through the cell membrane. Any particle that is too large to fit through the aquaporins or use the protein channels needs to enter through some form of membrane invagination. The membrane wraps around the particle and then pulls it into the cell. Pinocytosis is used for bringing in liquid particles and phagocytosis is used for bringing in solid particles. Receptor proteins sit on the surface of the cell membrane to detect the cell's surroundings, what types of things are trying to enter, and what is happening with other cells close by.

97. **How does active transport move molecules?**

[A] It moves them from an area of low pH to an area of high pH.

[B] It moves them in the direction of a higher osmotic potential of the cell.

[C] It moves them in a direction that is most likely to achieve equilibrium.

[D] It moves them from an area of high concentration to an area of low concentration.

[E] It moves them in the direction opposite to the one diffusion moves them.

The answer is E.

Active transport moves molecules from an area of low concentration to an area of high concentration. This is called going against the concentration gradient. Diffusion carries molecules down their gradient, from high concentration to low and does not take any energy. Active transport, on the other hand, does take energy because it is forcing more molecules into an area where there are already many.

98. **What would happen to a plant cell that loses a large concentration of its water?**

 [A] The cell will divide into smaller cells.

 [B] The cell will show a loss of turgor pressure.

 [C] The cell will release its waste products into the cytoplasm.

 [D] The cell will decrease in size equal to the amount of water loss.

 [E] The cell will keep its rigidity due to the cell membrane against the cell wall.

 The answer is B.

 Turgor pressure is the measurement of the amount of water within a plant cell. Turgor is caused when the vacuole within the cell inflates with water, which pushes the cell membrane against the cell wall. When a plant cell loses water, the vacuole shrinks, which causes the cell membrane to pull away from the cell wall, making the cell less rigid.

Questions 99 and 100 refer to the following information:

[A] Prophase I

[B] Metaphase I

[C] Metaphase II

[D] Telophase II

[E] Interphase

99. **During which phase of meiosis are adjacent pieces of homologous pairs going to trade places in order to increase the genetic diversity of the offspring?**

 The answer is A.

 During Prophase I, crossing over happens. This results in a small piece of one chromosome trading places with its homologue, which moves some of the genes on one chromosome to the other, and vice versa. This process increases the genetic diversity of the gamete so that if/when it gets fertilized, its genetics will be different enough from the parents' so that it has a better chance of survival.

100. **In oogenesis, when does one parent cell produce one cell with all of the genetic material and three "dead" polar bodies?**

 The answer is D.

 During telophase II, the egg cell divides again. This time, however, all of the genetic material has been placed in just one of the daughter cells. The other three cells (remember that in meiosis, each cell splits two times) do not have any genetic material. They get reabsorbed by the female's body. This is why human females (as most mammals) have a finite reproductive life. Each egg cell only produces one more. Therefore, it is possible to count the total number of eggs produced during her reproductive years.

101. **An organism was found to have DNA that contained 20% cytosine. Which of the following can be concluded about the DNA of this organism?**

 [A] It has 30% guanine.

 [B] It has 80% guanine.

 [C] It has 40% adenine and 40% thymine.

 [D] It has 20% thymine and 20% adenine.

 [E] It has 30% thymine and 30% adenine.

 The answer is E.

 According to Chargaff's rule, if an organism has 20% cytosine, then it also has to have 20% guanine. This means that the other two nitrogen bases must be equally divided among the other 60%, so 30% for thymine and 30% for adenine.

102. What can be concluded about the relationship between light intensity and the rate of photosynthesis, as seen in the graph?

[A] If this plant were given more light, the rate of photosynthesis will continue to increase.

[B] Photosynthesis is not dependent upon light intensity, but rather the temperature of the air.

[C] The rate of photosynthesis increases to a point of saturation regardless of how much light it is given.

[D] If light intensity were decreased, the rate of photosynthesis would continue to rise due to other external factors.

[E] An increasing light intensity can cause the rate of photosynthesis to fluctuate up to a certain point, and then cause it to drop off.

The answer is C.

The graph shows that the rate of photosynthesis and light intensity are related, but only to a point. As more light is provided, the rate increases. However, once it reaches a saturation point, it does not matter how much light is given. The chloroplasts cannot work any harder.

103. What is the function of aquaporins found the in the cell membrane?

 [A] They protect the inner parts of the cell.

 [B] They allow water to move back and forth.

 [C] They synthesize proteins from amino acids.

 [D] They package materials for export from the cell.

 [E] They eliminate cellular wastes from metabolism.

 The answer is B.

 Aquaporins are small pores found within the cell membrane especially designed to allow water to move back and forth across the membrane at will.

104. Gregor Mendel was an Austrian monk who studied the inheritance patterns of different characteristics of peas. He discovered that green coloring (G) is dominant to yellow (g) and that round seeds (R) are dominant to wrinkled (r). What would be the projected phenotypic ratio for a cross between two parents that were heterozygous for both traits?

 [A] All peas would be green and round.

 [B] All peas would be yellow and wrinkled.

 [C] Half of the peas would be green/round and half would be yellow/wrinkled.

 [D] There would be 9 green/round, 3 green/wrinkled, 3 yellow/round, 1 yellow/wrinkled.

 [E] There would be 1 green/round, 3 green/wrinkled, 3 yellow/round, 9 yellow/wrinkled.

 The answer is D.

 When doing a cross of this type it is important to separate the alleles properly so that each has the opportunity to be inherited. In this case, the alleles in the gametes would be GR, Gr, gR, gr. The Punnett square needs 16 possible offspring. All phenotypes will be represented in the ratio of 9:3:3:1.

105. Water is an essential molecule needed by all life on Earth. Which property of water makes it so useful?

[A] It is a polar ionic molecule.

[B] It is a polar covalent molecule.

[C] It is a nonpolar ionic molecule.

[D] It is a nonpolar metallic molecule.

[E] It is a nonpolar covalent molecule.

The answer is B.

Water is a polar covalent molecule. This means that the atoms that bond together to form it are sharing electrons in their outermost shells. It also means that there is a positive side and a negative side to the molecule after the atoms have bonded.

106. Which element was probably absent from the Earth's early atmosphere?

[A] Hydrogen (H^+)

[B] Oxygen (O_2)

[C] Sulfur (S)

[D] Carbon dioxide (CO_2)

[E] Carbon monoxide (CO)

The answer is B.

It is believed that the oxygen in the air came from photosynthesizing bacteria some 2.7 billion years ago. Prior to that time, there were no organisms around that could have performed the process to create the gas.

107. **What is the main difference between fermentation and glycolysis?**

 [A] Fermentation breaks down sugars, while glycolysis breaks down proteins.

 [B] Fermentation requires an anaerobic environment, while glycolysis requires an aerobic one.

 [C] Fermentation produces six molecules of ATP, while glycolysis produces only two.

 [D] Fermentation releases carbon dioxide into the atmosphere, while glycolysis releases oxygen.

 [E] Fermentation creates enormous stores of energy, while glycolysis uses large amounts of energy.

 The answer is B.

 Fermentation takes place in the absence of oxygen. For example, if you were making wine, you would place the grapes and yeast in an airtight barrel for the process to occur. Glycolysis needs oxygen in order to progress.

108. **During cellular respiration, which organelle is responsible for creating a concentration across its membranes?**

 [A] The nucleus

 [B] The mitochondria

 [C] The golgi apparatus

 [D] The nucleolus

 [E] The endoplasmic reticulum

 The answer is B.

 The mitochondria are the sites of most of the chemical reactions that occur during cellular respiration. This organelle has a double membrane system located within it that is used to create concentration gradients across which energy molecules can move.

109. **Which statement correctly identifies the effect of mutations on an organism's DNA?**

 [A] They are irreversible.

 [B] They are a source of variation.

 [C] They are only useful when found in germ cells.

 [D] They create selective pressures that cause evolution.

 [E] They benefit the species more than the organism.

 The answer is B.

 Mutations are changes found in an organism's DNA. These often harm the organism, but sometimes the mutation produces a trait that is beneficial. If the mutation happens to be found within a germ cell, then the new trait can be passed on. If the environment selects for it, over time, the entire population could show the new trait.

110. **Which of the following are present in both prokaryotic and eukaryotic cells?**

 [A] Cell wall, mitochondria, DNA

 [B] Mitochondria, ribosomes, RNA

 [C] Cell wall, DNA, nuclear membrane

 [D] DNA, RNA, ribosomes, cell membrane

 [E] Cell membrane, nuclear membrane, Golgi apparatus

 The answer is D.

 All cell types have genetic material (DNA and RNA) and ribosomes used to synthesize proteins. Prokaryotic cells lack nuclear membranes as well as other membrane-bound organelles. Not all eukaryotic cells have cell walls.

111. **In mammals, nerve cells have spaces between them, across which signals must be transferred. How do these signals get across this gap?**

[A] Neural impulses cause the release of chemicals that diffuse across the gap.

[B] Sodium and potassium rapidly flux back and forth to carry the signals across the gap.

[C] Electrical currents of varying voltages are emitted from one side of the gap to another.

[D] The calcium within the axons and dendrites of nerves adjacent to a gap acts as the messenger.

[E] Esterase serves as the messenger between nerve cells.

The answer is A.

When a signal reached a gap (called a synapse), it stimulates the release of chemicals called neurotransmitters. These chemicals then diffuse across the gap, carrying the signal with them. The neurotransmitters are then received by special receptors on the other side.

112. The following pedigree illustrates the pattern for an autosomal (non-sex-linked) gene that displays complete dominance. This pedigree only shows affected (black) and non-affected individuals.

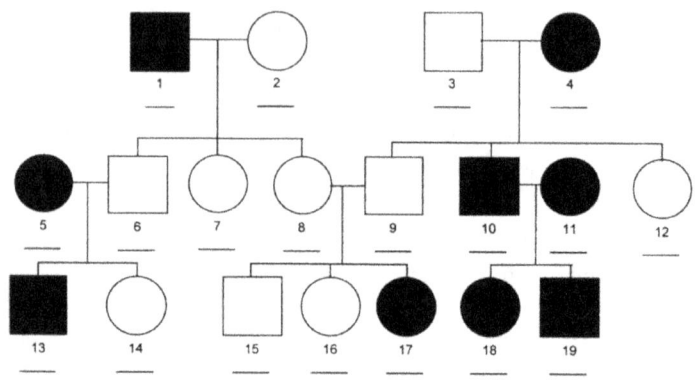

Which members of the above pedigree could be homozygous dominant?

[A] 2 and 3

[B] 2, 9, and 12

[C] 6, 7, and 8

[D] 2, 15, and 16

[E] 8, 9, and 15

The answer is D.

Person 2 is definitely homozygous dominant. This is clear because all of her children have the dominant phenotype even though their father was homozygous recessive. Persons 6, 7, 8, 9, and 12 are all heterozygous, since one parent was homozygous recessive and they have the dominant phenotype. Person 3 is heterozygous because half of the children have the recessive phenotype, so they must have gotten a recessive gene from their father. Persons 15 and 16 could be either heterozygous or homozygous dominant, since both of their parents were heterozygous.

113. The fossil record shows that large flying insects arose approximately 250 million years ago. Evidence also suggests that the oxygen concentration in the atmosphere at this time was much higher than it is now, approximately 28%. Scientists investigated the impact of oxygen concentrations on the growth of fruit flies.

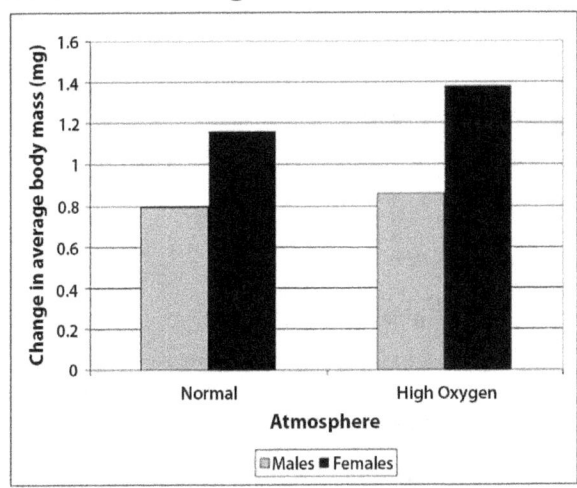

Based on the findings from the study shown in the graph, what would be a reasonable conclusion for the experiment?

[A] The increased sizes of insects were due to higher oxygen concentrations of the primitive atmosphere.

[B] The number of insect species was dependent on the existence of the higher oxygen concentrations.

[C] The metabolism of the large insects was the main reason ancient atmospheres had higher oxygen concentrations.

[D] The higher temperatures and oxygen concentrations contributed to the evolution of large animals on ancient Earth.

[E] Higher temperatures diluted the effects of higher oxygen on the body size of male insects.

The answer is A.

The data from the experiment show that higher oxygen concentrations contributed to a larger body mass of the fruit flies. Female fruit flies were especially affected, being 0.2 mg larger than those grown under normal atmospheric conditions.

114. **The worldwide population of the North Atlantic right whale has fallen so much the species is almost extinct. It also shows very little genetic diversity. Which of these is the most important risk factor the whales face?**

[A] The habitat of the whales is threatened by climate change and global warming.

[B] A dwindling food supply makes it more difficult for the right whales to survive and reproduce.

[C] Mutations are more likely to affect more individuals in populations with low genetic diversity.

[D] The low genetic diversity interferes with the ability of the population to respond to environmental changes.

[E] The small population makes it too difficult for whales to find mates.

The answer is D.

Since the population of whales is so low, it shows a very low genetic diversity. As a result, the gene pool for new traits is quite small. If a change in the environment occurred that required an adaptation to survive, it is unlikely the whales would be able to accommodate it.

115. **A scientist wants to find fossils from dinosaurs that lived at the end of the Jurassic period. Where would be a reasonable place to look?**

 [A] Igneous rocks from the late Triassic period (the era just preceding the Jurassic)

 [B] Sedimentary rocks from the late Triassic period (the era just preceding the Jurassic)

 [C] Igneous rocks from the early Cretaceous period (the era just after the Jurassic)

 [D] Sedimentary rocks from the early Cretaceous period (the era just after the Jurassic)

 [E] Metamorphic rocks from the late Triassic period (the era just preceding the Jurassic)

 The answer is D.

 Fossils would be expected in sedimentary rocks, which form from layers of sand/silt being deposited over time. Igneous rocks would rarely be expected to contain fossils since they form from molten rock. Sedimentary rocks from the late Triassic would already have formed by the time Jurassic organisms lived. Sedimentary rocks from the Cretaceous, however, would be expected to have formed around animals that died at the end of the Jurassic.

XAMonline
The CLEP Specialist
Individual Sample Tests in ebook format with full explanations

eBooks
All 33 CLEP sample tests are available as ebook downloads from retail websites such as **Amazon.com** and **Barnesandnoble.com**

American Government	9781607875130
American Literature	9781607875079
Analyzing and Interpreting Literature	9781607875086
Biology	9781607875222
Calculus	9781607875376
Chemistry	9781607875239
College Algebra	9781607875215
College Composition	9781607875109
College Composition Modular	9781607875437
College Mathematics	9781607875246
English Literature	9781607875093
Financial Accounting	9781607875383
French	9781607875123
German	9781607875369
History of the United States I	9781607875178
History of the United States II	9781607875185
Human Growth and Development	9781607875444
Humanities	9781607875147
Information Systems	9781607875390
Introduction to Educational Psychology	9781607875451
Introductory Business Law	9781607875420
Introductory Psychology	9781607875154
Introductory Sociology	9781607875352
Natural Sciences	9781607875253
Precalculus	9781607875345
Principles of Macroeconomics	9781607875406
Principles of Microeconomics	9781607875468
Principles of Marketing	9781607875475
Principles of Management	9781607875468
Social Sciences and History	9781607875161
Spanish	9781607875116
Western Civilization I	9781607875192
Western Civilization II	9781607875208

TO ORDER XAMonline.com or or

XAMonline
CLEP
Full Study Guides

CLEP College Algebra
ISBN: 9781607875598
Price: $34.95

CLEP Biology
ISBN: 9781607875314
Price: $34.95

CLEP Analyzing and Interpreting Literature
ISBN: 9781607875260
Price: $34.95

CLEP College Composition and Modular
ISBN: 9781607875277
Price: $19.99

CLEP College Mathematics
ISBN: 9781607875321
Price: $34.95

CLEP Psychology
ISBN: 9781607875291
Price: $34.95

CLEP Spanish
ISBN: 9781607875284
Price: $34.95

TO ORDER XAMonline.com or or

XAMonline
CLEP Subject Series
Collection by Topic
Sample Test Approach

CLEP Literature
ISBN: 9781607875833
Price: $34.95

CLEP Foreign Language
ISBN: 9781607875772
Price: $34.95

CLEP History
ISBN: 9781607875789
Price: $34.95

CLEP Sociology
ISBN: 9781607875796
Price: $34.95

CLEP Science
ISBN: 9781607875802
Price: $34.95

CLEP Mathematics
ISBN: 9781607875819
Price: $34.95

CLEP Business
ISBN: 9781607875826
Price: $34.95

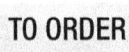

 or or

XAMonline
CLEP Favorites
Collection by Topic
Sample Test Approach

CLEP Five Favorites
ISBN: 9781607875765
Price: $24.95

CLEP Military Favorites
ISBN: 9781607875512
Price: $24.95

TO ORDER or or **BARNES & NOBLE** BOOKSELLERS

XAMonline.com

www.ingramcontent.com/pod-product-compliance
Lightning Source LLC
Chambersburg PA
CBHW060938230426
43665CB00015B/1987